CW01272722

People's CAR

A facsimile of
B.I.O.S. FINAL REPORT NO. 998 INVESTIGATION INTO THE DESIGN
AND PERFORMANCE OF THE VOLKSWAGEN OR GERMAN PEOPLE'S CAR

First published in 1947

Reprinted with a new introduction by
Karl E. Ludvigsen

London: The Stationery Office

© Crown Copyright Reserved, 1947. Published under licence
from the Controller of Her Majesty's Stationery Office
Introduction copyright Karl E. Ludvigsen 1996

ISBN 0 11 290555 2

Any applications for reproduction should be made to
the Controller of Her Majesty's Stationery Office, The Copyright Unit
St Clements House, 1–16 Colegate, Norwich NR3 13Q

British Library Cataloguing in Publication Data

A CIP Catalogue record for this book is available from the British Library

The Stationery Office

Published by The Stationery Office and available from:

The Publications Centre
(mail, telephone and fax orders only)
PO Box 276, London SW8 5DT
General enquiries 0171 873 0011
Telephone orders 0171 873 9090
Fax orders 0171 873 8200

The Stationery Office Bookshops
49 High Holborn, London WC1V 6HB
(counter service and fax orders only)
Fax 0171 831 1326
68–69 Bull Street, Birmingham B4 6AD
0121 236 9696   Fax 0121 236 9699
33 Wine Street, Bristol BS1 2BQ
0117 9264306 Fax 0117 9294515
9–21 Princess Street, Manchester M60 8AS
0161 834 7201 Fax 0161 833 0634
16 Arthur Street, Belfast BT1 4GD
01232 238451 Fax 01232 235401
The Stationery Office Oriel Bookshop
The Friary, Cardiff CF1 4AA
01222 395548 Fax 01222 384347
71 Lothian Road, Edinburgh EH3 9AZ
(counter service only)

Customers in Scotland may
mail, telephone or fax their orders to:
Scottish Publications Sales
South Gyle Crescent, Edinburgh EH12 9EB
0131 479 3141 Fax 0131 479 3142

Accredited Agents
(see Yellow Pages)

and through good booksellers

Printed for The Stationery Office by Premier Print, Nottingham
Dd 302661 C25 12/96

Introduction
by Karl E. Ludvigsen

You will never see the documents that change the course of the car industry. They are presented to boards or management committees, debated, decided upon and then locked away – perhaps only to be seen by the eyes of later generations. You will never see them – *with one exception*: this historic technical assessment of the performance and the post-war potential of the car that struck fear into the hearts of the German auto industry and all who owned her: the Volkswagen.

Conceived as a car that would be made cooperatively by the German auto companies to absorb their excess capacity in the Great Depression, the Volkswagen metamorphosed with relentless energy into a government-backed project that threatened to swamp all its rivals, not only at home but also abroad, with its high-volume production and high-pressure weekly payment plan (see Appendix, p.136).

The numbers involved were mind-boggling by European and indeed American standards at the time. The definitive form of the Volkswagen-to-be was revealed on 26 May 1938 when the cornerstone of the factory at Fallersleben was laid by Chancellor and *Führer* A. Hitler. Production was expected to start in 1939 with 100,000 being produced in 1940, twice as many in 1941 and then up to 450,000 per year in the first stage of the plant's development, with a workforce of 17,500 on two shifts.

That was to be just the beginning. The huge, modern plant on the Mittelland Canal was designed by Peter Koller to be expanded so 30,000 workers could build 800,000 to one million Volkswagens yearly. More than half would be exported to earn valuable currencies for the New Germany.

Even in the midst of Germany's aggressive rearmament the factory and its peculiar rear-engined car, designed as its project Type 60 by the Porsche organisation, would get first choice of tools, machinery, materials, supplies and personnel. No wonder rivals in Germany, France, Italy and Britain lost sleep over the immense potential for harm the project promised them. The car's only saving grace, they assured each other, was that it was 'not only unconventional but ugly in the extreme'.

Ultimately the VW plant would indeed reach and even exceed those ambitious volume goals. Post-war rivals would continue to

Prof. Ferdinand Porsche with a KdF prototype on the Grossglockner Pass road, 1939. (*Ludvigsen Library*)

count beetles, not sheep, while they tossed and turned. But the 1939–40 plans had to give way to reality. In 1941, when construction of the plant at Wolfsburg (as its town became known) was halted, no more than half of the first stage was completed, and that only in shell form, not yet equipped to make cars of any kind.

Before the war began in earnest only 210 Volkswagens of the final design had been built, hand-tooled by Daimler-Benz as show and display vehicles and toys for the Nazi elite. Instead of their mooted price of RM990, equivalent then to $396 or some £85 (about as much as a good 350 cc motorcycle), the few cars made were sold initially for RM8,000 apiece. This fell to RM3,000 by 1944 for the tiny handful – 630 cars – of civil-type VWs made in the plant.

Not discovered by the Allied bombers until the first of five raids on 8 April 1944, the huge plant complex east of Hanover was equipped by the Germans to produce parts for aircraft, V-1 flying bombs – and military versions of the Volkswagen. In the late 1930s the Porsche office, which was jointly responsible for running the plant, developed the square-rigged, high-riding Type 82 that became known as the *Kübelwagen* or 'bucket-car'. An amphibian version, the *Schwimmwagen* or 'swimming-car', entered production in 1942. In its final form it was Porsche's Type 166.

The first KdF Wagens built in 1941 at Wolfsburg. (*Ludvigsen Library*)

Parts for the vehicles were sourced throughout Greater Germany. The Peugeot plant at Sochaux, which became a subsidiary of the Volkswagen works in June 1943, supplied 20,000 connecting rods, 3,000 flywheels and 5,000 raw forgings for crankshafts – the last a rush order, we learn from a December 1944 C.I.O.S. report. Type 82 body panels were made by Ambi-Budd in Berlin and shipped west to Wolfsburg for assembly.

Of the swimming beetles, 14,283 were built during the war. Of the Type 82 'bucket-cars', output was 50,435 following the production of the first of these 'German Jeeps' in 1940. Equipped with large-section 'aero' tyres that let them take advantage of their light weight, the Type 82s proved their merit by skimming over the desert sands of North Africa. Field-Marshal Erwin Rommel said his *'Kübel'* could go wherever a camel could. And if one did get stuck in a ditch it was easy enough for a few soldiers to heave its light chassis out and send it buzzing onward.

Able though they were, the Type 82s could do little without fuel. Some were simply left behind with empty tanks when the rest of Rommel's army withdrew after the British counter-attacked successfully in North Africa late in 1942. Many examples, captured intact, were commandeered by the Allied forces, who valued their agility. And several were liberated for study back home. One went to the US Army's Aberdeen Proving Grounds. The Americans, impressed by the agility of the *Kübelwagen*, 700 pounds lighter than their Jeep, described it in detail in their technical manual TM E9-803.

Another Type 82, its frame members still stuffed with 100 pounds of Saharan sand, was decanted in the British Midlands in the cold of January 1943 at the headquarters of Humber Ltd. Built in 1941, it carried chassis number 1339 and Ambi-Budd body number 1777. Bracketry evidenced that it had been used to carry a welding kit to repair its fellow warriors. It is the vehicle that you find described in Part I of this report by the British Intelligence Objectives Sub-committee (B.I.O.S.).

Top: A Schwimmwagen emerges from the water showing the boat-like profile of the design. *Above:* A rear view shows the propeller in the lowered position ready for entering the water. (*Ludvigsen Library*)

For its capacity to study the technologies of the vanquished Germans the B.I.O.S. drew upon the skills of various UK government ministries, all of which were represented on a sub-committee chaired by the Board of Trade. This decentralised structure was no hindrance to the huge scope of the research conducted under the auspices of the B.I.O.S., which resulted in more than 2,000 reports issued and published by HMSO.

Motor vehicles were the responsibility of the Ministry of Supply, which allocated the *Kübelwagen* to Humber for study. In addition to being one of the marques of the Rootes Group, acquired by the Rootes family interests in 1931, Humber was also the car-producing unit in the Group. Hillmans and, later, Sunbeam–Talbots as well as Humbers were engineered and built by Humber under the technical direction of Bernard 'B.B' Winter.

A dispatch rider in the Great War, Winter enjoyed the confidence of arch-salesman Billy Rootes and his administrator brother Geoffrey. Rising through the service side of the Rootes organisation from a service engineering post, he became head of service before becoming technical chief of Humber. Bernard Winter was an advocate of the progressive development of the *status quo*; innovation was not an objective, as his car designs well illustrated.

Winter's team at Humber provided a meticulous and well-illustrated description and discussion of the Type 82 as they found it (pp. 4–64). Their report, with illustrations, was summarised over 11 pages of the July, 1944 issue of *Automobile Engineer*. Humber highlighted the use of lightweight magnesium for the camshaft gear, blower rotor, oil pump housing and engine crankcase (although it is described as 'aluminium' in the text). They noted that the engine-cooling fan also drew cooling air through the dynamo.

The drive train justly deserved their special attention. Humber noted the neat ring of nine pins that slid to engage the constant-mesh third and fourth gears. It describes in detail the patented ZF cam-type differential that gave the Type 82 such grip in slippery going in spite of having only two-wheel drive. It showed how hub gearing was ingeniously used both to raise ground clearance and lower final drive ratio – technology later used to advantage in the VW Kombi or Transporter.

A simple 'idiot light' warned of low oil pressure. The tubular dampers at the front worked only on rebound, when they were compressed. The wheels were cheaply and neatly attached to the brake drums by five studs. Lucas assisted with the detailed study of the electrical system, including the distance-indicating lamp at the rear that allowed a following driver to judge his distance at either 50 or 140 metres according to the perceived merging of illuminated windows.

Part II of the report consists of a performance comparison by Humber between a civil version of the VW, built under British supervision at Wolfsburg after the war (called Type 11 using the British designation) and received by Humber on 18 February 1946, and a Hillman Minx. Interestingly the comparison Minx was not the pre-war Mark II but a prototype of the post-war Mark III model. A dimensional and performance check gives support to the assumption that this did not yet have the envelope body of the final Mark III Minx, introduced in September 1948, 'but was already equipped with that model's independent front suspension.

The car tested by Humber against the VW would have looked much like this Mark II Hillman Minx, albeit with prototype independent front suspension. (*Ludvigsen Library*)

Baldly Humber reports that the subject VW was down a good 25 per cent in compression pressure in one cylinder and 8 per cent in another, both on the right-hand bank, 'but no action was taken to dismantle the engine to find the cause'. As well, Humber seemed to take the view that the VW should be penalised for its lighter weight by being loaded more heavily than the Minx for the tests!

Described as 'definitely poor' by Humber, the VW's dismal braking performance is shown in a table on page 75. Eighty pounds of pressure on the pedal (quite a lot) generated only half the deceleration of the (in this case) more-heavily-loaded Minx. According to Ferry Porsche, son of the designer, the cable-operated brakes were an economy measure. Not only were they cheaper to make; they also avoided any royalty payment to Lockheed, which held the key patents on hydraulic brakes. The brakes, admitted Ferry in his autobiography, 'were, in fact, a feature about which my father had always had misgivings'.

Carrying equal loads, each with a driver and passenger, the VW at 1,861 pounds yielded much better fuel consumption than the 2,324-pound Minx prototype. At average speeds of 37.6 and 36.0 mph respectively, the VW and Minx returned 43.0 and 34.6 miles per Imperial gallon, in that order. Even when burdened with five passengers and 160 pounds of luggage-equivalent the plucky VW managed 40.0 mpg at 36.3 mph. Humber admitted that no more than four people could squeeze into the Minx.

With the help of the Pressed Steel Company Ltd, in Part III of the report Humber assessed the body design and structure of a Type 11 VW saloon. It produced an excellent set of drawings of details of the body as built by VW under its British controllers. The assessors gave it as their 'considered opinion that from the Body Engineering point of view the design of this vehicle is exceptionally good, and shows a great advance on previous constructional methods'.

Points of particular interest on the body included the integral heating and demisting system, with its ducts built into the body structure, and the facia panel designed to suit both right-hand- and left-hand-drive versions. Humber found that the design reduced welding as much as possible by stamping 'large complete panels, often of a complex type', and joining many large panels by wrapping their edges over instead of welding them.

Attention was drawn to the light yet stiff design of the front and rear lids. Features that we came to take for granted much later, such as pull-out door handles and an automatic self-latching stay for the front deck lid, were part of the design. Such details as the stamped parts of the door latches and window regulators attracted positive comment in the construction of what was called 'an exceptionally rigid and semi-unitary vehicle'.

Parts IV and V of the report are, it must be vouchsafed, peculiar contributions to the assessment by Singer Motors and AC Cars, respectively. Both drove what was evidently a *Kübelwagen* of post-war construction, called by them a Type 21 although the designation used by the British was 'Type 51' for this model.

That quality control was not yet up to VW's later standards was clear from the faults reported by Singer, which included a 'collapsed' wheel bearing and inoperative wipers, horn and speedometer. Singer found the engine, gearbox and ZF differential 'extremely noisy' but the roadholding 'extremely good', with positive steering and little pitch and roll. The clutch and gear shifting were praised.

Faults that transformed themselves into virtues after a few miles of driving the 'Type 21' were features of the brief assessment by AC Cars. The instant response of the inflexible power train, although initially disconcerting, soon won admirers. So too did the unfamiliar closeness of the driver to the front of the car. 'Soon overcome with a little practice', AC said, 'the excellent forward visibility is then appreciated'.

The AC assessor seemed to pay the Porsche designers a compliment by concluding 'that this vehicle is a useful education for those who think there is a future for the rear engined car', a type all but unknown in Britain – *pace* Sir Charles Dennistoun Burney and his cars. Added AC, 'The case for all round independent suspension is clearly demonstrated to be a valuable feature.' AC itself, in the event, was one of the last British car makers to cease putting solid axles under the fronts of its cars.

A 1,131 cc engine from this *Kübelwagen* was transported to Dagenham. There Ford revealed its secrets on one of its dynamometers and recorded them on the 8th of February, 1946. At a measured 5.63:1 the compression ratio was lower than the nominal 5.80:1, but the engine's peak power of 24.3 bhp at 3,250 rpm was about what the VW's designers would have expected. Ford's placing of the torque peak at 2,000 rpm was also in accord with VW's tests.

Unfamiliar as they were with air-cooled engines, the Ford engineers were notably disconcerted by the fact that the cooling air emerging from the left-hand cylinders (right-hand as they were looking at the engine) was some 20 to 40 degrees Farenheit higher than the right-side exhaust air, depending on the load (p. 129). This was certainly caused by the fact that the air on that side passed through the oil cooler, thereby taking away additional heat. What was important to the cylinders, of course, was the *difference* in temperature that they experienced, not the absolute level.

Ford considered that the flat-four had an excessive thirst for fuel and tried different carburetter jettings to reduce the richness of the mixture. (Solex comments on the carburetter in Part VII.) This brought a 16 per cent reduction in specific fuel consumption. Stating that 'the carburetion could be improved to give a more economic fuel consumption', Ford concluded its report by saying, 'The performance is not outstanding and the general noise in operation is excessive, the engine could not be recommended *in its present form* [emphasis added]'.

In spite of Ford's dusty view of the engine, it will be evident from the above that the assessors found much to admire and indeed to emulate in the VW's radical and ruthlessly simplified design. The saloon's body construction was rife with advanced and clever simplifications and cost reductions. The VW was substantially lighter than its nearest

British counterpart and yielded much better fuel consumption. It shifted easily, handled well and gave a good ride.

How, then, are we to make sense of the General Observations volunteered by Humber Engineering on page 10, observations that have often been quoted as sounding the last post for any wide acceptance of the VW vehicle among the victorious powers?

Astonishingly, Humber was unwilling to grant that the VW design offered any advantage in weight! It criticised its engine for not producing higher specific power, ignoring the fact that it was intentionally designed as a *Drosselmotor*, deliberately throttled below its power potential to ensure its durability on Germany's new Autobahns.

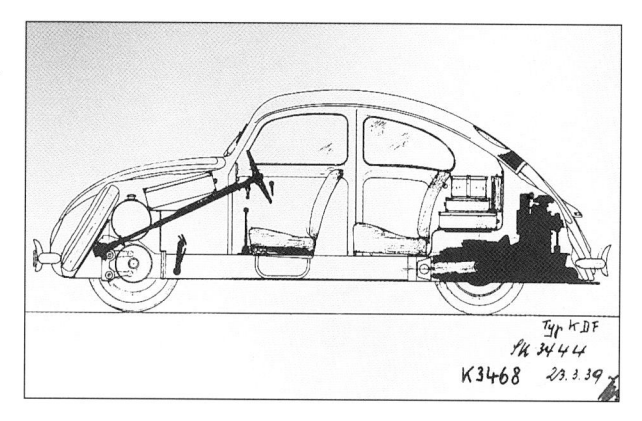

Most devastating to the later perception of this report was the final paragraph of Humber's General Observations:

Porsche office drawing of the Type 60 in its final form, the KdF Wagen. (*Ludvigsen Library*)

Looking at the general picture, we do not consider that the design represents any special brilliance, apart from certain of the detail points, and it is suggested that it is not to be regarded as an example of first class modern design to be copied by the British industry.

To be sure, Humber said that these remarks 'should be treated purely as their views'. But both at the time and later this dismissive verdict by the Bernard Winter team tended to obscure the many merits of the VW design as related by the individual assessors.

For an industry that was obsessed by refined running, that viewed the Rolls-Royce as 'The Best Car in the World', the rough and raucous wartime Volkswagens must have come as a shock. Even in the first post-war cars their virtues were presented in a very crude, coarse and, in the case of the Type 82 dismantled by Humber, worn-out form. Bouncing along in a bellowing 'bucket-car', innocent of sound damping of any kind, some British engineers were not impressed.

AC's assessor, however, seemed to grasp the situation: 'From the general construction one gets the impression that the designer has given just enough but no more, therefore as a war vehicle this is no doubt acceptable, but as a civilian vehicle considerable modification would be required to conform to the standard expected.' Bearing in mind that he was speaking about the open military version of the VW, this was incontrovertible.

Heinz Nordhoff, who in 1947 was asked by the British to manage the VW plant, later told Gordon Wilkins that 'when I assumed the responsibility of putting this wrecked factory into high volume production, the car itself was still full of "bugs". It really was what you call an ugly duckling.' Tough enough, cheap to build, using only the materials available in wartime, Porsche's radical light-car concept now needed taming for peacetime.

Summarising and reviewing the reports on the German industry in B.I.O.S. Overall Report No. 21, published in 1949, senior General Motors engineer Maurice Olley reflected perceptively as follows:

Although a number of the reports give space to the Volkswagen, and one report [this one] is devoted exclusively to it, one doubts whether they convey a true idea of the possibilities of this car. They refer chiefly to captured military equipment, or to staff cars hastily 'run up' from pre-war stocks for the military government, and subjected to military usage. We have yet to see what the makers, left to themselves, can make of this car in the way of a vehicle suitable for normal civilian use. Perhaps the critics of the vehicle forget that large areas of the world are still looking for cheap transportation, and that the Model T Ford, which started world motorization, also had technical faults.

Humber's summing-up also vouchsafed the view that 'little or no special advantage has been obtained in production cost', in spite of evidence to the contrary, especially in the body design. To be sure, the assessment did not deal with the VW's production system. This was the responsibility of other B.I.O.S. survey teams.

One such team (Final Report No. 768) stated that, 'Compared with other automobile factories in Germany, and visualising the originally intended factory layout, the Volkswagen effort is outstanding and is the nearest approach to production as we know it.' They also reported that 'the condition of the plant and equipment is good, and could operate satisfactorily for many years'.

These were the views of a team from Ford, one of the few companies in the world that knew something about car production on the massive six-figure scale foreseen for the Wolfsburg factory. In contrast, the total production capacity of Humber before the war, for all its marques, was only 50,000 units per annum, a minor role in an industry that made some 350,000 cars each year. Its conservative engineers would have had extreme difficulty in projecting their thinking to VW's intended volumes, which were an order of magnitude greater.

But not *everyone* at Humber missed the point of the Volkswagen. In the fuel-rationed 1940s one of its senior engineers, Craig Miller, built a prototype of a small car nicknamed 'Little Jim'. Among its features was a rear-mounted air-cooled flat-twin engine! Like Porsche, Miller had clearly concluded that, at the time, this was the low-cost solution for small-car design.

And in 1955 none other than Bernard Winter startled young Humber engineers Mike Parkes and Tim Fry when he green-lighted their plan to develop a small two-seater car. Like Miller, but unaware of his project, they started out with an air-cooled flat twin (built by Villiers) mounted at the rear of their 'Slug' prototype. When the car became a four-seater its need for power grew; the result was the rear-engined Hillman Imp, launched in 1963. Humber had been susceptible to heterodoxy after all!

Meanwhile under its British Army administration the Wolfsburg factory bestirred itself into car production. British Forces themselves needed wheels to patrol their sector of a divided Germany; that was the source of the initial demand. In 1945 output was 1,785 units and in 1946 a remarkable 10,020. After the British and American occupation zones were fused in September, 1946, GIs could buy beetles through their Post Exchange stores for a heady $645.

Production was chiefly the Type 11 saloons; the supply of body panels to make the *Kübelwagen* dried up after the Soviets dismantled the Ambi-Budd plant in Berlin and transported it back to Mother Russia. They had an ingenious proposal for the VW plant as well: a simple redrawing of the sector boundaries would put the factory, about five miles away, within the Russian zone! To their credit the British spurned this suggestion, which would have seen yet another factory looted of its equipment.

Assembly facilities for the VW at Wolfsburg in October 1948 were still provisional at best. (*Ludvigsen Library*)

Rootes Motors was offered the VW plant as war reparations, either to run *in situ* or to transplant to the UK. Clearly influenced by his engineers, Billy Rootes not only refused the plant but also openly ridiculed its product. In 1945 the French, who would later imprison Ferdinand Porsche, came close to acquiring and re-siting the factory; again the British resisted. Australian auto pioneer L.J. Hartnett spent three weeks in 1946 assessing the plant as a possible acquisition for his nation but concluded that commercially it would be too risky to build the VW there in the high volumes that its tooling justified and indeed demanded.

Senior Vauxhall engineer Maurice Platt later admitted that he had shared in the general post-war view that the VW was not only an ugly duckling but also a dead duck: 'My own view, at the time, was that it was noisy, uncomfortable and tricky to handle; a crash gearbox and mechanically operated four-wheel brakes (requiring a considerable pedal effort) added to the driver's difficulties. Of all the pre-war cars available for resurrection, this seemed to be one of the least promising.'

But when in 1947 the Overseas Operations Division of General Motors began thinking of building an 'austerity model' abroad to meet third-world requirements, one of its young engineers, William Swallow, said that instead of adapting its own components GM should acquire the Volkswagen, which would be ideal for such conditions. Swallow had been a development engineer at Pressed Steel, where the body of the Type 11 saloon was assessed in such positive terms for this report.

Recalled Maurice Platt, 'His report was studied by the engineering hierarchy in Detroit with the predictable result that his recommendation was dismissed. Detroit was as critical of the primitive Volkswagen as Luton or Coventry and was certain that it could not survive for more than a year or two, even in underdeveloped countries, in competition with the postwar American cars that would soon come into production. Bill nearly lost his job through having shown such poor judgement.' Reprieved by GM, Bill Swallow later became the knighted chairman of Vauxhall.

Another young executive, Henry Ford II, weighed up Volkswagen in early 1948 during an exploratory trip to Europe. Although he did not visit the plant he met the newly-installed Heinz Nordhoff at Frankfurt Airport and discussed with the British authorities the idea of Ford acquiring a 51 per cent share in the car-making enterprise. 'This was a hope, a dream, and not a plan', wrote Ford historians Allan Nevins and Frank Ernest Hill.

Would VW, under Ford's control, have grown into the business that built the car that ultimately displaced Ford's Model T as the most-produced auto in history? We are obliged to doubt it.

In September 1949, just a decade after the start of World War II, the British military authorities signed over control of the plant and its products to the Federal German government and the state of Lower Saxony. The British signatory was Colonel C.R. Radclyffe, in charge of all vehicle production in the British Zone. His deputy for Lower Saxony was Major John P.C. Macgregor. In charge of the plant itself had been Major Ivan Hirst. Commander Richard H. Berryman had been responsible for production.

Colonel C.R. Radclyffe (centre) with Heinz Nordhoff (second left) talks to Ludwig Ehrhard (right) at the 1949 ceremonies which handed over the Volkswagen Works to the West Germans. (*Ludvigsen Library*)

This British report, negative as it was about the VW vehicle itself, helped keep the Volkswagen Works in the hands of Germany. For that reason it is by any standards historic. But without the affection for the car shown by the British military at Wolfsburg, and their determination to help it survive, there would have been no Volkswagen Works for Germany to be given.

'Had this handful of British officers not engaged themselves so daringly for the factory and its workers,' wrote Walter H. Nelson in *Small Wonder*, 'there could never have been a VW today. Inevitably, the factory would have decayed or been dismantled. Despite their dedication, hard work and self-sacrifice, the German Volkswagen executives there at the time would have been powerless to save the plant in those years, had they stood alone. Fortunately they did not . . .' Fortunately indeed.

B.I.O.S. FINAL REPORT No. 998

ITEM No. 19

INVESTIGATION INTO THE DESIGN AND PERFORMANCE OF THE VOLKSWAGEN OR GERMAN PEOPLE'S CAR

This report is issued with the warning that, if the subject matter should be protected by British Patents or Patent applications, this publication cannot be held to give any protection against action for infringement.

BRITISH INTELLIGENCE OBJECTIVES SUB-COMMITTEE

LONDON—H.M. STATIONERY OFFICE

Price 12s. 0d. net

Your attention is drawn to the existence of B.I.O.S. Information Section, 37, Bryanston Square, London, W.1., to which enquiries about German **Technical** Processes, etc., covered in B.I.O.S. Reports, and requests for German Technical Information otherwise available in the U.K., can be sent.

THE SOCIETY OF MOTOR MANUFACTURERS AND TRADERS
LIMITED.

INVESTIGATION INTO THE DESIGN AND PERFORMANCE OF
THE VOLKSWAGEN OR GERMAN PEOPLE'S CAR.

Reported by:-

A. C. CARS LTD.

FORD MOTOR CO. LTD.

HUMBER LTD.

SINGER MOTORS LTD.

SOLEX LTD.

BIOS Target Number : 19/9.

BRITISH INTELLIGENCE OBJECTIVES SUB-COMMITTEE.
32 Bryanston Square, W.1.

Der Innenlenker, Preis RM 990.- ab Werk Fallersleben

PREFACE

This report consists of results of tests and inspections by a number of vehicle or component manufacturers and gives as near as possible a complete picture of the performance and design of the Volkswagen, or German People's Car.

The car, designed chiefly by Dr. Porsche, was about to go into production as a civilian car in 1939, but as a wartime measure a modified vehicle (type 82) was produced for the German Army. A similar vehicle (type 21) with an enlarged engine of 1131 cc. and 5.25 - 16 tyres was in production in 1945, and as a temporary measure a few vehicles were produced having a saloon body on the military chassis, and was known as type 51. In 1946 the civilian saloon (type 11) was again in production having the 1131 cc. engine and 5.00 - 16 tyres. The main points of difference between the military and civilian vehicles are the raised chassis and Z.F. limited-slip differential on the former, whereas the latter has a normal differential and lower chassis. The chassis is raised by a modification to the stub axle assembly at the front, and the super-imposition of a spur gear reduction (15 teeth to 21 teeth) at the axle ends at the rear.

Whilst this report is basically concerned with the technical aspects of design, it is however important to consider the reason for the production of this vehicle and the conditions making it possible. The two illustrations shown on the frontispiece are taken from the official German catalogue announcing the vehicle and its method of purchase by special savings stamps. A translation of the rules governing the purchase is given in the appendix. The price given in this catalogue is RM. 990, and the rate of exchange at the close of 1938 was £1 = RM. 11.65. A far better rate was available for travellers visiting Germany and might have been applicable to cars purchased for export, had the authorities wished to capture the export trade.

---ooOoo---

CONTENTS.

Part 1.   Design report by Humber Ltd., on the military vehicle type 82.   4

Part 11. Comparative road performance test by Humber Ltd., on the civilian vehicle type 11.   65

Part 111. Body engineering report by Humber Ltd., on the civilian vehicle type 11.   77

Part 1V. Road performance test by Singer Motors Ltd., on the military vehicle type 21.   114

Part V.   General impression by A.C.Cars Ltd., on the military vehicle type 21.   117

Part V1. Bench test by Ford Motor Co. Ltd., on the 1132 c.c. engine.   119

Part V11. Report by Solex Ltd., on the carburettor fitted to type 82.   134

Appendix.   136

PART I

REPORT ON EXAMINATION

OF

GERMAN LIGHT AID DETACHMENT VEHICLE
TYPE V.W. 82

"VOLKSWAGEN"

Produced by

THE ROOTES GROUP

Engineering Dept., Humber Ltd.

CONTENTS

	PAGE
INTRODUCTION	8
GENERAL DATA OF COMPLETE VEHICLE	12
ENGINE AND MOUNTING	14
CONSTRUCTIONAL DETAILS OF ENGINE	18
ENGINE DATA	22
CLUTCH	24
GEARBOX	24
GEARBOX DATA	26
REAR AXLE DRIVE	27
REAR AXLE DRIVE DATA	30
REAR SUSPENSION	32
REAR SUSPENSION DATA	34
FRONT SUSPENSION, SPEEDOMETER DRIVE AND STEERING HEAD	35
FRONT SUSPENSION DATA	38
SHOCK ABSORBERS	39
STEERING	42
STEERING DATA	43
BRAKING SYSTEM	45
CONTROLS	46
BRAKE DATA	47
WHEELS AND TYRES	48
BODY-CONSTRUCTIONAL DETAILS	49
GENERAL DATA	52
ELECTRICAL EQUIPMENT	53

LIST OF PLATES

Plate No.	Subject	Page
1	COMPLETE VEHICLE, THREEQUARTER FRONT VIEW	7
2	,, ,, SIDE VIEW	7
3	,, ,, THREEQUARTER REAR VIEW	7
4	CHASSIS	13
5	ENGINE, THREEQUARTER REAR VIEW	14
6	,, THREEQUARTER FRONT VIEW	17
7	ENGINE COMPARTMENT	17
8	CRANKCASE, L.H. HALF, EXTERIOR	18
9	,, ,, ,, INTERIOR	18
10	,, R.H. ,, INTERIOR	18
11	CRANKSHAFT AND GEAR WITH CONNECTING RODS ASSEMBLED	19
12	PISTON AND GUDGEON PIN	19
13	CAMSHAFT AND GEAR ASSEMBLY, ALSO GEAR WHEEL	19
14	CYLINDER HEAD SHOWING COMBUSTION CHAMBER	20
15	CYLINDER HEAD SHOWING VALVE ROCKER GEAR	20
16	VALVE PUSH ROD AND TUBE	20
17	OIL PUMP	21
18	DYNAMO PULLEY CONSTRUCTION	21
19	AIR CLEANER, EXPLODED VIEW	22
20	CLUTCH UNIT	24
21	GEARBOX AND GEARS (ASSEMBLED IN L.H. INTERIOR CASING)	25
22	GEARBOX, SHOWING SELECTOR CONTROL AND REVERSE WHEEL (L.H. INTERIOR CASING)	25
23	GEARBOX CASING, R.H. INTERIOR	25
24	REAR AXLE SHAFT END AND CYLINDRICAL SEGMENT FOR UNIVERSAL JOINT	27
25	REAR AXLE REDUCTION GEAR CASING AND BRAKE BACK PLATE	27
26	REAR AXLE TUBE AND SHAFT COMPLETE	27
27	DIFFERENTIAL AND AXLE SHAFT ASSEMBLY	28
28	DIFFERENTIAL, EXPLODED VIEW	28
29	REAR SUSPENSION, R.H. WHEEL	32
30	FRONT SUSPENSION UNIT	35
31	STEERING HEAD, R.H. EXPLODED VIEW	35
32	FRONT SHOCK ABSORBER, EXTERIOR	39
33	REAR SHOCK ABSORBER, EXPLODED VIEW	40
34	STEERING UNIT, EXPLODED VIEW	42
35	BRAKE SHOE ASSEMBLY, L.H.	45
36	HANDBRAKE, EXPLODED VIEW	46
37	PEDAL UNIT	46
38	DISTRIBUTOR AUTOMATIC ADVANCE MECHANISM	53
39	DISTRIBUTOR, SHOWING CONTACT BREAKER	53
40	DISTRIBUTOR DRIVING SHAFT	53
41	DISTRIBUTOR, SHOWING H.T. WIRES	54
42	DYNAMO BODY, SHOWING FIELD WINDING, BRUSH HOLDER AND REGULATOR	55
43	DYNAMO ARMATURE AND COMMUTATOR ASSEMBLY	55
44	STARTER BODY, SHOWING FIELD WINDINGS AND BRUSH HOLDER	56
45	STARTER ARMATURE AND COMMUTATOR	56
46	SOLENOID EXTERIOR, WITH COVER REMOVED	57
47	BATTERY, SHOWING CELL REMOVED	57
48	INTERIOR, SHOWING FACIA PANEL	59
49	FACIA PANEL, FRONT VIEW, SHOWING FUSE BOX COVER REMOVED	59
50	FACIA PANEL, REAR VIEW, SHOWING WIRING	59
51	HEADLAMP	61
52	NOTEK HEADLAMP	61
53	HELLA SPOT LIGHT	62
54	NOTEK DISTANCE INDICATOR, REAR LAMP AND STOP LIGHT	62
55	ELECTRIC HORN, EXTERIOR	63
56	ELECTRIC HORN, INTERIOR	63
57	WINDSCREEN WIPER, EXPLODED VIEW	63
58	WINDSCREEN WIPER, REAR COVER REMOVED	64

LIST OF FIGURES

Figure No.	Subject	Page
1	BODY, SIDE ELEVATION	9
2	,, PLAN	9
3	,, FRONT VIEW	9
4	,, REAR VIEW	9
5	CHASSIS, SIDE ELEVATION	11
6	,, PLAN	11
7	,, FRONT VIEW	11
8	ENGINE CROSS SECTION, END	15
9	,, ,, ,, PLAN	15
10	,, LONGITUDINAL SECTION	16
11	,, OILING DIAGRAM	16
12	INLET VALVE LIFT	20
13	EXHAUST VALVE LIFT	20
14	ENGINE AIR COOLING SYSTEM, PHANTOM VIEW	21
15	GEARBOX AND CLUTCH (LONGITUDINAL SECTION)	24
16	DIFFERENTIAL, PHANTOM VIEW	28
17	,, DEVELOPMENT DRAWING	29
18	REAR AXLE AND GEARBOX (CROSS SECTION)	31
19	REAR SUSPENSION, MIN. GROUND CLEARANCE	33
20	,, ,, MAX. ROLL POSITION	33
21	,, ,, WHEEL MOVEMENT	33
22	,, ,, R.H. SIDE, PLAN	34
23	,, ,, LONGITUDINAL LINK, ACTION DIAGRAM	34
24	FRONT SUSPENSION, MAX. ROLL POSITION	36
25	,, ,, WHEEL MOVEMENT	36
26	,, ,, L.H. SIDE, FRONT VIEW	37
27	,, ,, L.H. SIDE, PLAN VIEW	37
28	,, ,, EXPLODED VIEW	37
29	FRONT SHOCK ABSORBER, WORK DIAGRAM	39
30	,, ,, ,, SECTIONAL VIEW	39
31	REAR SHOCK ABSORBER VALVE, SECTIONAL VIEW	40
32	,, ,, ,, ,, TEMPERATURE DIAGRAM	41
33	,, ,, ,, ,, WORK DIAGRAM	41
34	STEERING BOX AND COLUMN, SECTIONAL VIEW	43
35	BRAKE SHOE OPERATION	45
36	FOOTBRAKE AND HANDBRAKE CONTROLS	47
37	BRAKE CONTROL HEAD	47
38	WHEEL, CROSS SECTION	48
39	BODY DETAILS	50
40	,, ,, SECTIONAL VIEWS	51
41	UNDERFRAME, TOP	52
42	,, UNDERSIDE	52
43	DISTRIBUTOR DISTRIBUTION CURVE	54
44	,, TEMPERATURE RISE	55
45	,, OUTPUT	55
46	,, CIRCUIT	56
47	SOLENOID, SECTION	57
48	,, CONNECTIONS	57
49	,, CHARACTERISTIC OF PLUNGER RETURN SPRING	57
50	,, CHARACTERISTIC OF CONTACT SPRING	57
51	WIRING DIAGRAM	60
52	NOTEK HEADLAMP, HORIZONTAL BEAM DISTRIBUTION	61
53	,, ,, VERTICAL BEAM DISTRIBUTION	62
54	WIPER ARM AND SPINDLE SECTION	63

Plate 1.—Complete vehicle, threequarter front view

Plate 2.—Complete vehicle, side view

Plate 3.—Complete vehicle, threequarter rear view

INTRODUCTION

In January, 1943, a German light aid detachment vehicle (Volkswagen) was received by Humber Limited, who were instructed to make a complete examination and prepare a technical report. The vehicle was captured in the Middle East and it was ascertained that it was originally fitted with a gas welding kit for dealing with repairs to military vehicles.

Military Conversion of the Volkswagen Compared with the Original Design of People's Car

The vehicle under examination is based on the Volkswagen (German People's Car) and from the available descriptive matter the military version differs in the following respects.

The peacetime Saloon body is replaced by an entirely new open type tourer body which has obviously been designed to suit its military role.

Hub reduction gears have been added.

Special tyre equipment and wheels are also used.

Features of Noteworthy Interest

The engine is fitted at the rear of the vehicle, behind the axle.

A horizontally opposed, four-cylinder, air-cooled type of engine is employed.

The engine air cooling system incorporates a rotor and cowling arranged to circulate air to the cylinders and also to an oil cooler.

Independent wheel suspension is provided for all wheels, torsion bars being used. The front torsion bars are of unusual and ingenious design to obtain soft springing in a compact form. The front suspension, complete with the steering unit, shock absorbers and track rod system, forms a very compact assembly unit.

The speedometer drive is housed neatly within one of the stub axles.

Extensive use is made of aluminium and magnesium base alloys, and a very good finish imparted to the die castings.

Plain carbon steels are used in preference to alloy steels, except in special cases such as valves, etc. Nickel has not been used for the manufacture of any of the parts, and copper has been added to the cast iron components in order to produce a similar effect to nickel. The hardening elements used in the steels are manganese, chromium and molybdenum.

A special dog-type gear engagement is incorporated in the gearbox, using steel rods in grooves; this is fully described in the text. The design of differential is ingenious, having only a partial slip, thus obviating wheel spin and therefore very suitable for cross country and muddy conditions. This is patented under German Pat. Spec. No. 639876, and British Pat. Spec. No. 431020, both patents taken out by Gottfried Weidmann.

Hub reduction gearing is used to obtain a lower overall ratio by a simple conversion of the original design of Volkswagen, and this also gives the increased ground clearance required for traversing across country.

Chassis consists of a light gauge pressed steel underframe; this is arranged in a " back-bone " construction and also provides the floor. A comparatively strong chassis, especially torsionally.

The location of the rear suspension swing arms or struts above the axle enables a cheap and light form of strut to be used according to claims stated in Patent No. 544748 F. Porsche. Ground clearance is also increased by raising the torsion bar relatively.

The body was of open tourer type fitted with a collapsible fabric hood and provided with a steel trunk designed to carry the welding plant. It is thought that a study of the sections and methods of construction described in this report will be valuable.

In order to ascertain full particulars of the design, technical data, weights and dimensions, the vehicle was dismantled and assembly drawings, together with a detail description of the design and construction, prepared. Photographs (Plates 1, 2, 3 and 4) were taken on receipt of the vehicle and show its general appearance when it arrived at these works.

From the condition of the vehicle as received for examination it was apparent that it had covered a considerable mileage—unfortunately the speedometer was not functioning and the exact mileage could not therefore be verified.

The following identification plates were fitted inside the engine compartment at the rear of the body:—

Identification Plate relating to the Chassis.

Volkswagen Wks. Ltd. (People's Car Factory Ltd.)

Type VW.82.

Engine capacity 985 cu. cms.

Year of manufacture, 1941.

Weight of vehicle (unladen) 685 kilograms (1,510 lbs.).

Weight of vehicle (max. laden weight) 1,175 kilograms (2,590 lbs.).

Weight on Front Axle (max. laden weight) 450 kilograms (992 lbs.).

Weight on Rear Axle (max. laden weight) 724 kilograms (1,598 lbs.).

Chassis Number 001339.

Identification Plate relating to the Body.

Ambi Budd Pressworks, Johannesthal, Berlin. No. 1777. Year 1941.

Identification Plate relating to the Electrical Equipment.

Screened, based on Group III Bosch.

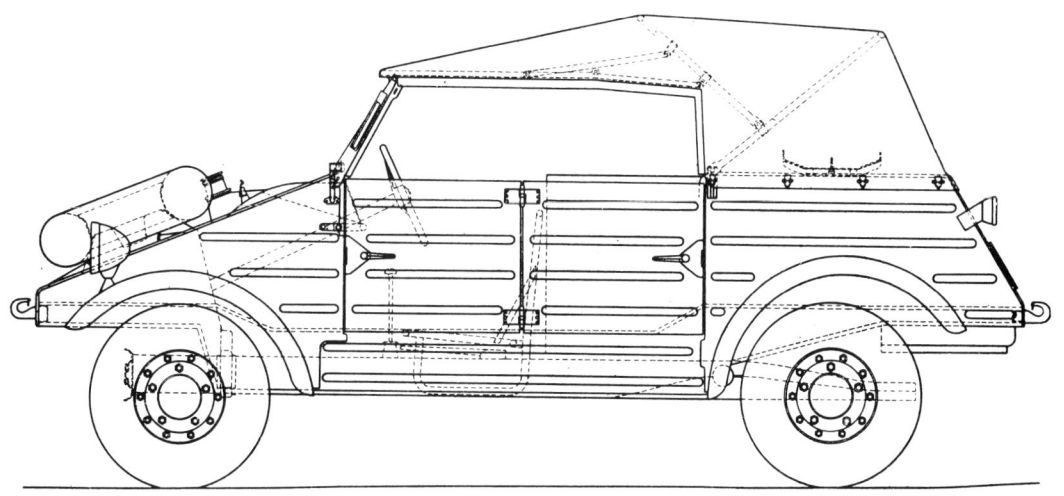

Fig. 1.—Body, side elevation

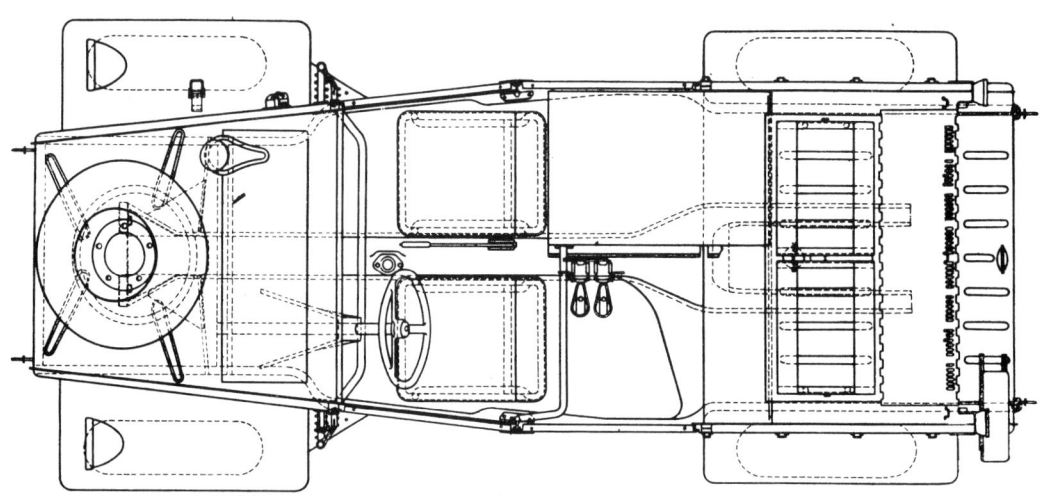

Fig. 2.—Body, plan

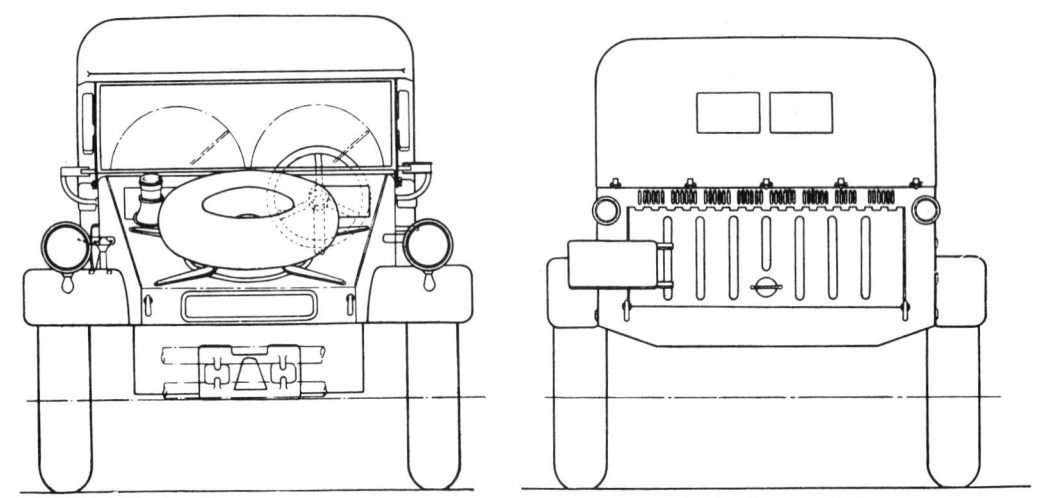

Fig. 3.—Body, front view

Fig. 4.—Body, rear view

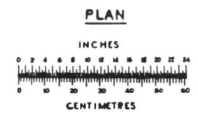

PLAN

The actual weights of the vehicle as received were as follows, which it will be noted are in excess of those given on the identification plate.

Actual weight of vehicle (unladen) 14 cwts. 3 qrs. (1,652 lbs.).

Actual weight on Front Axle (unladen) 5 cwts. 3 qrs. (644 lbs.).

Actual weight on Rear Axle (unladen) 9 cwts. 0 qrs. (1,008 lbs.)

General Observations

The following general observations are made by Humber Engineering and should be treated purely as their views.

The design is particularly interesting because it is quite uninfluenced by any previous traditions, and it is doubtful if the question of whether the public would or would not like a car with an air-cooled engine positioned at the rear was considered by the designer. This model has departed almost entirely from the conventional motor-car and features of interest have already been referred to above.

In spite of the assumed freedom of the designer and the unconventional vehicle produced, little or no special advantage has been obtained in production cost, neither does it appear that any improvement in performance or weight compared with the more conventional type of vehicle known in this country has been achieved.

So far as materials are concerned, no signs of the use of any ingeniously applied materials have been found, in other words the material specification is, with few exceptions, very parallel with what is already well known in this country. The use of plastics is not apparent. The tyres are, however, manufactured from synthetic rubber.

A study of the engine indicated that the unit was, in certain details, most inefficient. The design of the inlet manifold makes it clear that the designer did not intend the unit to produce power proportionate to its capacity, and from a study of both the design and condition of the crank bearings it is very doubtful whether it was even capable of giving reliable service had it produced a performance commensurate with its size.

Looking at the general picture, we do not consider that the design represents any special brilliance, apart from certain of the detail points, and it is suggested that it is not to be regarded as an example of first class modern design to be copied by the British industry.

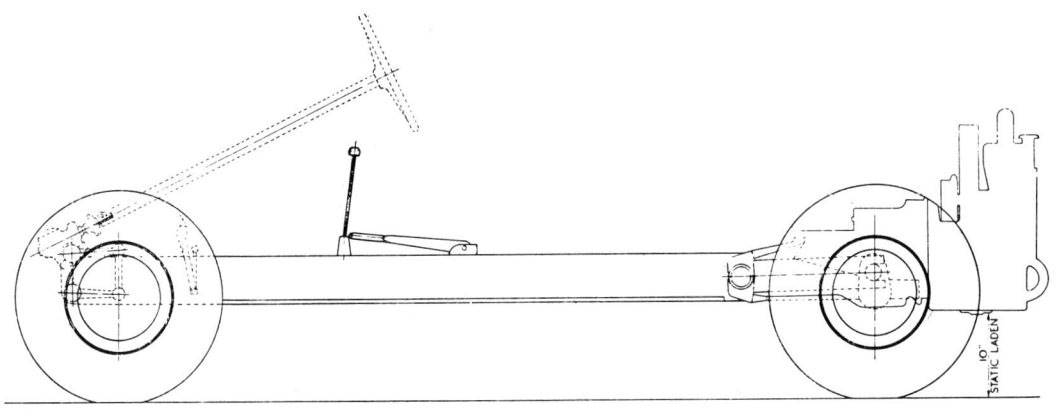

Fig. 5.—Chassis, side elevation

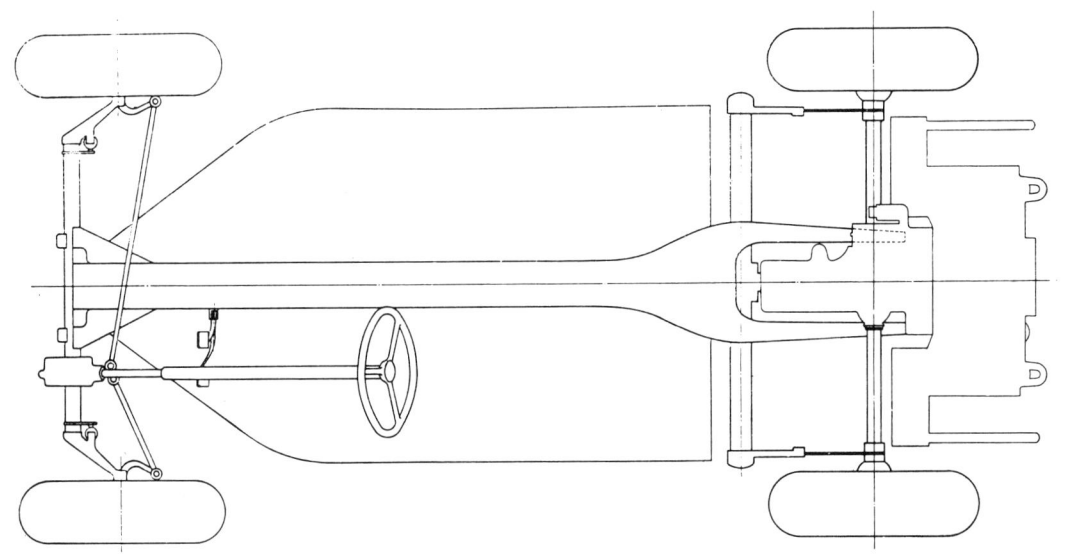

Fig. 6.—Chassis, plan

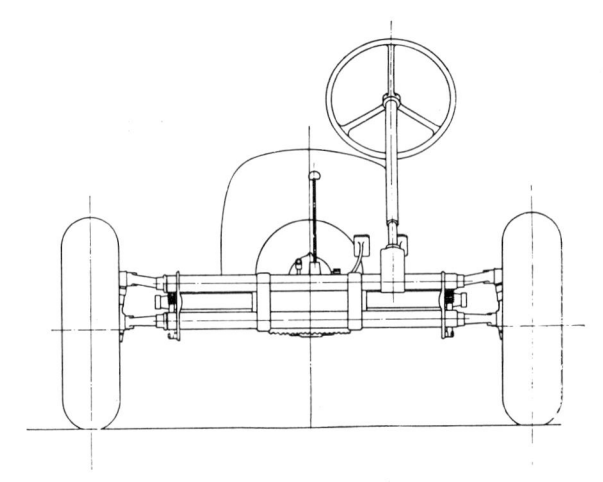

Fig. 7.—Chassis, front view

GENERAL DATA OF COMPLETE VEHICLE

Dimensions

Chassis No. 001339.

Overall length : 147·5 ins. (12 ft. 3½ ins.).

Overall width : 63 ins. (5 ft. 3 ins.).

Overall height (top of hood) : 63·5 ins. (5 ft. 3½ ins.)

Wheelbase, normal, static laden condition : 94 ins. (7 ft. 10 ins.).

Wheelbase variations under pitching conditions :—

(a) Front Wheel at full rebound position and Rear Wheel at full bump position : 95·40 ins., 5 ins. engine clearance.

(b) Front Wheel at full bump position and Rear Wheel at full rebound position : 93·40 ins.

Track : Front Wheels : 54·625 ins. (4 ft. 6⅝ ins.).

Rear Wheels : 55·125 ins. (4 ft. 7⅛ ins.) laden.

Turning circle. R.H. lock : 30 ft. 5 ins.
L.H. lock : 36 ft. 8 ins.

Weights

Chassis : 7 cwts. 0 qrs. 21 lbs.

Rear Axle : 4 cwts. 3 qrs. 21 lbs.

Front Axle : 2 cwts. 1 qr. 0 lbs.

Engine

Engine No. 001346.

Make : Volkswagen.

Type : Four-cylinder, air-cooled, horizontally opposed, overhead valves.

Capacity : 0·985 litres (60 cu. ins.).

R.A.C. Rating : 12·2 H.P.

Air Cleaner : Oil Bath type.

Clutch

Make : Fichtel & Sachs — Komet.

Type : K.10 Single Plate Dry Clutch.

Size : 180 mm. (7·09 ins.) O/Dia.

Gearbox

Type : Constant mesh helical—Top and third speeds. Straight Spur—First and Second Speeds. Special dog engagement on Top and Third Speeds. Gear engagement on First and Second Speeds.

Control : Remote control. Ball change type gear lever.

Ratios. Four forward speeds and one reverse.
Top : 0·8 : 1.
Third : 1·25 : 1.
Second : 2·07 : 1.
First : 3·6 : 1.
Reverse : 6·6 : 1.

Rear Axle

Type : Enclosed swinging half axle. Spiral bevel drive and final hub reduction gear.

Axle ratios : Spiral Bevel : 4·43 : 1.
Hub gears : 1·40 : 1.
Overall ratio : 6·2 : 1.

Transmission Unit Weight

Gearbox, Clutch, Rear Axle, Reduction Gear and Starter less Brake Shoes and Drum : 147 lbs. weight.

Overall Ratios

Fourth speed : 4·96 : 1.
Third speed : 7·75 : 1.
Second speed : 12·83 : 1.
First speed : 22·32 : 1.
Reverse speed : 40·92 : 1.

Suspension

Type : Independent wheel springing on all wheels.

Longitudinal link type on front wheels. Swinging half axle and longitudinal arm on rear wheels.

Springs : Front torsion bar ; rectangular section ; multi-blades.

Rear torsion bar ; round section.

Shock Absorbers

Front : Hydraulic, direct, single-acting telescopic type.

Rear : Hydraulic, piston-operated, double-acting type and lever arm.

Wheels and Tyres

Make of Tyre : Continental.

Size of Tyre : 690 × 200 (8 × 12) smooth tread, aeroplane type.

Size of Wheel : 4·25 × 12, flat base rim.

Steering

Make : Volkswagen steering box.

Type : Worm and rocker shaft (segment nut interposed).

Number of turns of handwheel (lock to lock): 2·75

Connections : Divided track rod, directly coupled.

Brakes

Make : Volkswagen.

Type : Internal expanding; two shoes; floating operation.

Control : Cable operated, non-compensated type.

Petrol Tank

Capacity : 9 gallons.

Electrical Equipment

Dynamo : Bosch type, RED 130-6 2600 AL.89.
6 volt. Ventilated.
Speed : 1·75 times engine speed.

Starter : Bosch type, EEDD 4-6 L3P 6 volt. Screw push type.

Drive : Pinion, 9 teeth ; Ring, 109 teeth.

Ignition : Bosch ignition, coil type TL6. Bosch distributor, type VE4BS276. Screened.

Plate 4.—Chassis

ENGINE AND MOUNTING

GENERAL DESCRIPTION

The engine was originally designed for the "Volkswagen," which was exhibited at the International Motor Show held in Berlin in 1939, the manufacturers being Volkswagen, Fallersleben, near Hanover, Germany.

Makers' identification marks: No. 001346 (stamped on the crankcase); No. 0402 (cast on each of the cylinder heads).

Other markings: Firing order, 1, 4, 3, 2 (cast on the crankcase), No. 1 Cylinder being nearest the flywheel.

The complete unit includes a sheet metal cowling mounted above the engine, and incorporating a blower: this circulates air for cooling the cylinders and the oiling system. The blower consists of a rotor, mounted on one end of the dynamo armature shaft, which is driven at the opposite end by a "V" belt drive from the crankshaft.

The engine is of the overhead valve, horizontally opposed four-cylinder type, consisting of two banks, each bank having two cylinders which are separately cast and interchangeable. Detachable cylinder heads of aluminium silicon alloy are fitted; these are cast in pairs and located in the cylinders by means of spigots formed on the latter. Both these are secured to the crankcase by long studs, screwed direct into the crankcase, the cylinder head joint being formed between the top face of the cylinder spigot and the head. Bronze alloy valve seat inserts, phosphor bronze valve guides, and steel sparking plug inserts are employed.

The crankshaft is supported by three main bearings, and an additional bearing which acts as a steady for the auxiliary drives; the thrust loads are taken by the bearing nearest to the flywheel. The crankshaft main journals consist of thick steel shells lined with lead bronze; all the bearings are exceptionally narrow, especially the centre one, which is split for assembly purposes.

The connecting rods are made from steel stampings, a relatively thick layer of bearing metal being run direct into the big end. The bolts securing the connecting rod caps have hexagon socket-type fittings on the heads.

The aluminium crankcase is made in two halves, and is split on the vertical centre line through the main bearings; the halves are secured together by means of bolts and studs. An oil sump is formed integral with the crankcase; the underside is generously finned. In addition, the casing serves as a mounting for the various accessories such as the dynamo, oil cooler, blower equipment, etc.

A single camshaft driven at half engine speed by single helical gears from the crankshaft runs direct in the aluminium crankcase, and actuates the overhead valves through push rods, each cam operating two rods. The whole of the valve gear is pressure lubricated.

The distributor is mounted on top of the crankcase, and is driven by spiral gears from the rear end of the crankshaft. The driven shaft consists of the spindle, and gear and cam for operating a petrol pump, and is made from a steel stamping, which is hardened and ground. It is supported at both ends and runs directly in the crankcase, the gear end thrust also acting against the crankcase facing.

The petrol pump is an AC diaphragm type, mechanically operated, mounted on the left-hand of the crankcase, on a neat moulding which also houses the operating rod.

Plate 5.—Engine, threequarter rear view

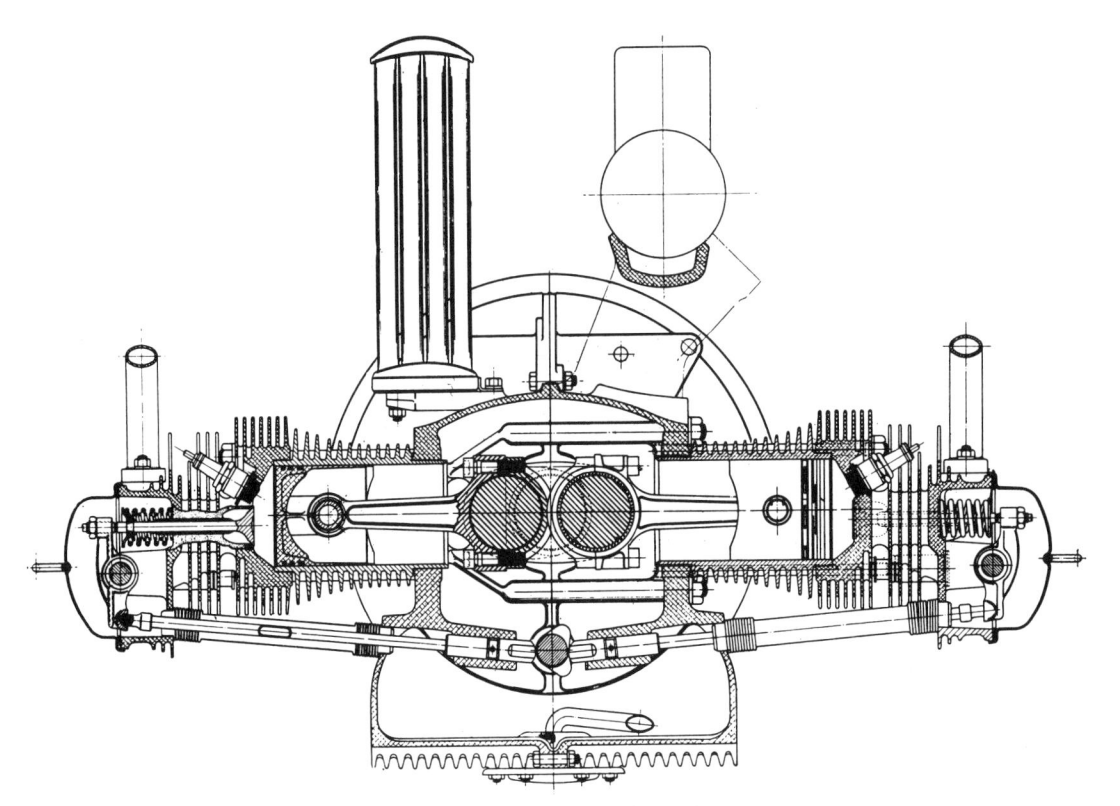

Fig. 8.—Engine cross section, end

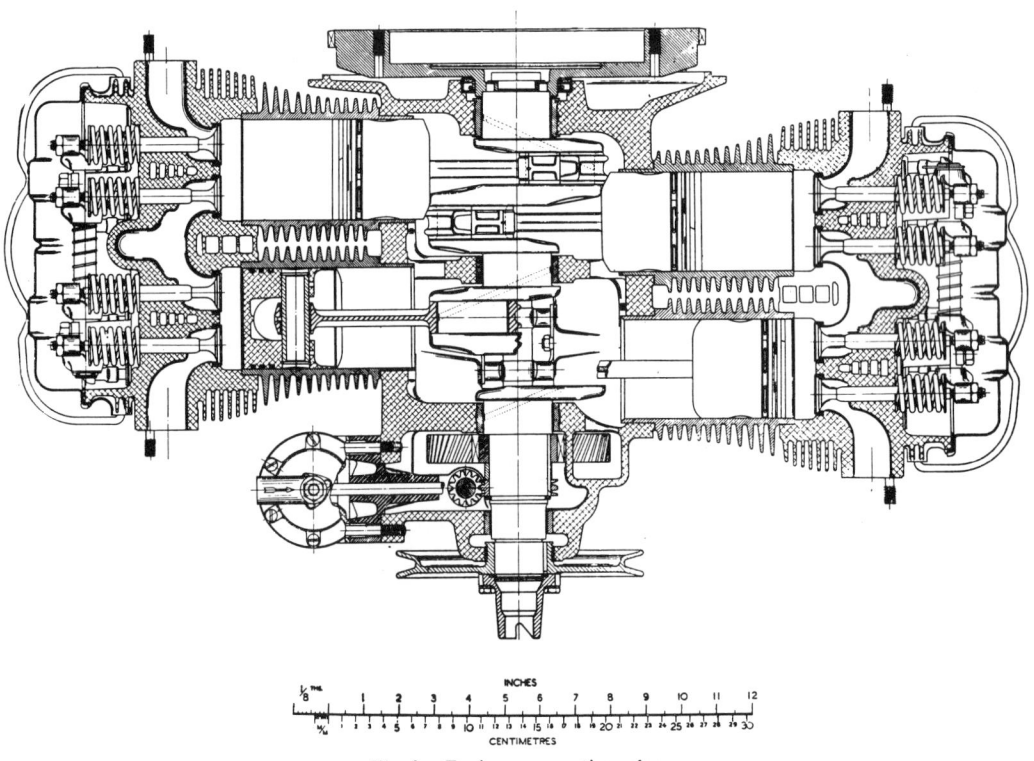

Fig. 9.—Engine cross section, plan

15

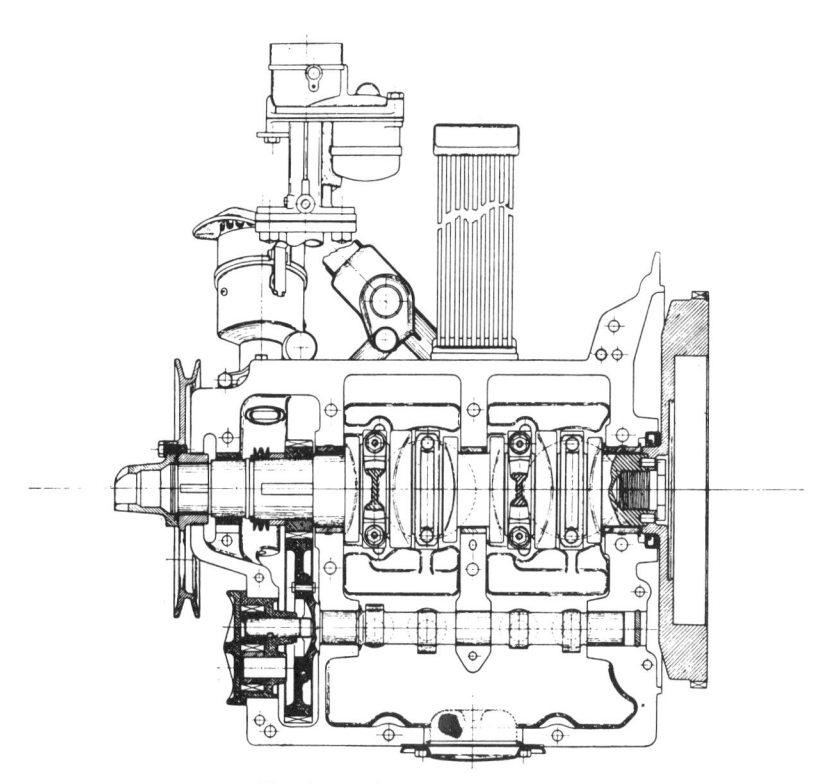

Fig. 10.—Engine, longitudinal section

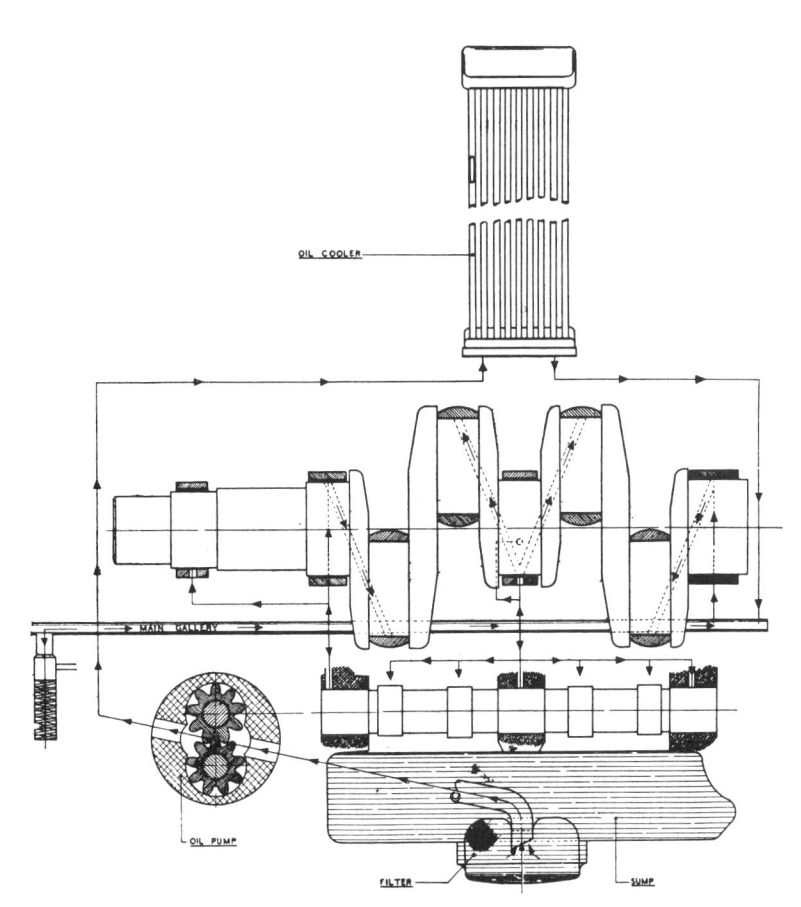

Fig. 11.—Engine, oiling diagram

Plate 6.—Engine, threequarter front view

A single downdraught "Solex" carburettor is fitted. This is connected to the cylinder head by an extremely small bore inlet pipe, which has a central hot spot. The latter is obtained as the result of pressure pulsation which causes the exhaust gases to flow past the hot spot.

The crankcase has an extension which provides a saddle mounting for the dynamo and also forms a convenient oil filler orifice on account of its hollow construction. The dynamo is driven by means of a "V"-shape belt, adjustment being provided.

The flywheel is a steel stamping with integral starting gear teeth, and is spigoted to the end of the crankshaft and driven by four dowels. It is secured to the crankshaft by a single centre bolt, the latter being hollow so as to include a self-lubricating bush for the clutch shaft. No locking device is provided.

Condition

The main bearings were scored, badly worn and of oval shape. The connecting rod big ends were worn ovally, but the small ends were round. There was considerable end play in the crankshaft. Valves were quite a good fit in the guides. The camshaft and its bearings were in good condition. The tappet push rod guide holes were in good condition. The connecting rod in No. 3 cylinder was bent towards one end of the engine. The pistons were in good condition. The compression rings and one scraper ring had broken and these had been replaced with new rings.

Plate 7.—Engine compartment

CONSTRUCTIONAL DETAILS OF ENGINE

CRANKCASE

The crankcase is an aluminium alloy sand casting, built in two halves, the joint passing vertically through the centre lines of both the main bearings and the camshaft bearing—the camshaft itself running directly in the crankcase. An oil sump is formed integral with the crankcase, of fairly wide and shallow proportions so as to afford maximum ground clearance, fins being cast in a longitudinal direction on the underside. (Provision is made for a detachable gauze filter which is centrally mounted on the base of the sump.)

Plate 8.—Crankcase, L.H. half, exterior

A large diameter flange formed at the flywheel end is utilised as an engine mounting face with spigot fitting. Lubrication is through drilled oil ways in the crankcase. Platforms

Plate 9.—Crankcase, L.H. half, interior

on top of the crankcase carry the oil cooler, dynamo and blower equipment.

Plate 10.—Crankcase, R.H. half, interior

CRANKSHAFT

The crankshaft is a steel stamping, hardened and ground on all bearing surfaces, and supported on four main bearings. Markings indicate that the crankshaft was statically balanced only. Oil ways are drilled for pressure feed lubrication from the main journal to the connecting rod big end bearings. It is understood that work on cast crankshafts was in progress prior to hostilities.

CRANKSHAFT MAIN BEARINGS

The main bearings consist of lead bronze metal linings on thick steel shells; all the bearings are formed in the shape of a continuous ring, with the exception of the centre bearing which is split for assembly purposes. It would appear that the thick shells are introduced to give additional rigidity to the bearing, especially as these are supported in the aluminium crankcase.

CONNECTING RODS

Very short and comparatively stiff H section steel connecting rods are used; the big end bearings consist of babbit metal of thickness ·06-in. min., ·10-in. max., run directly in the rod and cap. A phosphor bronze bush is pressed into the small end. The connecting rod cap has two vertical webs for stiffness and the securing bolts are screwed into the caps, this being the only means of location. Hexagon socket type bolt heads are used.

PISTONS

Flat head aluminium alloy die cast pistons are used, having a generous amount of metal above the gudgeon pin bosses but very thin, unsplit, skirts. Two compression rings and one scraper ring are positioned at the top end.

Plate 11.—Crankshaft and gear with connecting rods assembled

Fully floating gudgeon pins are fitted, **and** retained in the piston by means of spring **steel** wire rings fitted into grooves.

Plate 12.—Piston and gudgeon pin

CAMSHAFT

The camshaft is of cast iron, supported on three bearings, and has four cams, each of which operates two push rods and valves A flange cast at one end provides an attachment for the gear drive; this end is also slotted for engagement with a tongue on the oil pump spindle. Single helical teeth are cut on the gear wheel (of magnesium base alloy material) the wheel being riveted to the camshaft flange. End-thrust on the camshaft is absorbed between two flanges at the driven end.

CYLINDER HEADS

The detachable cylinder heads are sand cast in pairs in aluminium silicon alloy. A recess encloses the valve gear, and the whole casting is generously finned. The valves are arranged in line (disposed horizontally), two valves being fitted for each cylinder. The two inlet valves are situated in the centre and in consequence the inlet ports are interconnected in the cylinder head casting. Seating inserts are of bronze and are either pressed or shrunk in position and retained by peening over the surrounding metal; the valve guides are of phosphor bronze. Steel inserts for the sparking plugs are fitted and these are screwed up to a flange formed on the inserts and pegged in position to prevent them from working loose.

VALVE GEAR

Two valves per cylinder are fitted, the exhaust and inlet valves being identical. The valve springs are also interchangeable, two springs being fitted to each valve. The overhead valve gear is contained in a chamber cast integral with the cylinder head and is totally enclosed by a pressed steel cover which is held in position by a spring steel

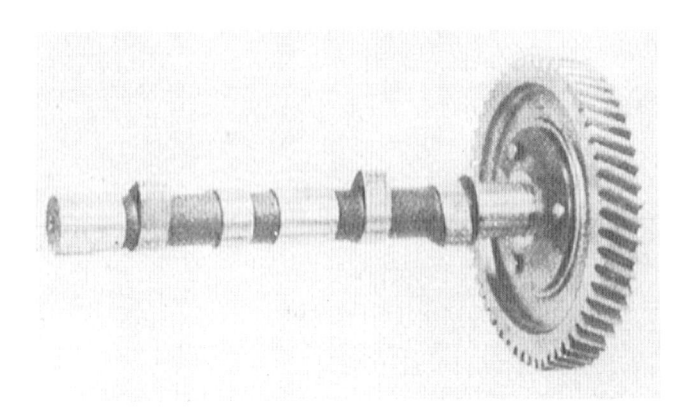

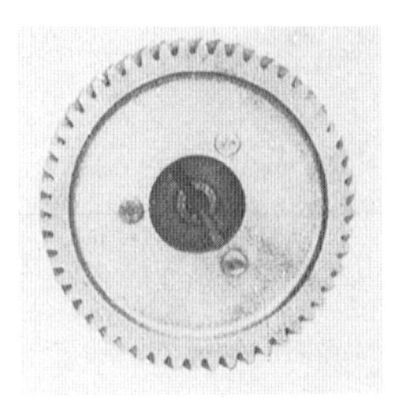

Plate 13.—Camshaft and gear assembly, also gear wheel

19

Plate 14.—Cylinder head showing combustion chamber

Plate 16.—Valve push rod and tube

wire clip. The rocker arms fulcrum direct on to hardened and ground shafts, an oilway being drilled from the push rod spherical seating to the rocker shaft. Adjustment for setting valve clearance is provided by a pin screwed into the rocker arm and operating the valve. Push rods operate the valve gear and these are unusual in their design, being in effect composite tappets and rods; the tappet ends of the push rods are guided (as a bearing), in reamed holes in the crankcase. The push rods are of long tubular construction in aluminium base alloy, both the tappet end and the opposite end (which is spherical) being hardened and ground steel components. It will be observed that this design causes slight bending to occur in the push rod whilst operating the valve, due to the path of the rocker arm spherical end. The tappet is held against rotation, and has a radius formed at the base, at an angle to compensate for the push rod angle. Welded steel tubes enclose the push rods, and these have a series of bellows formed at each end, to act as a spring loaded seal, allow for misalignment, and assist in the machining operations of the holes in which the push rod tube fits. See Fig. 12 for Valve Lift curves (the lift was measured at the valve, and therefore includes valve gear deflection and wear).

OIL SYSTEM

The engine lubricating oil is contained in a sump formed in the crankcase, and including a detachable gauze filter. Oil is drawn from the filtered compartment through the suction tube to a gear type oil pump, housed away from the sump in the rear end of the crankcase. The pump is driven by a tongue engaging in a slot on the end of the camshaft, the pump being partly below the oil level in the sump. There are nine straight spur gear teeth on each wheel, the driving wheel being fixed to the spindle by arc process—a somewhat unusual application. The pump body is

Plate 15.—Cylinder head showing valve rocker gear

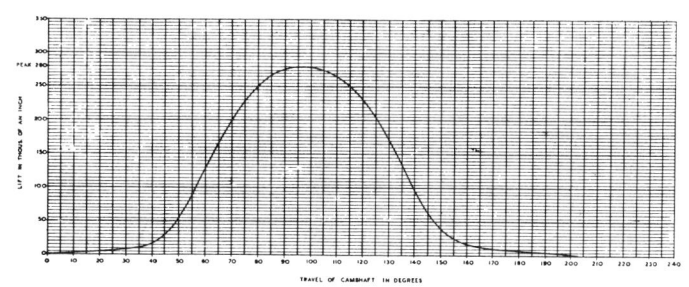

Fig. 12.—Inlet valve lift

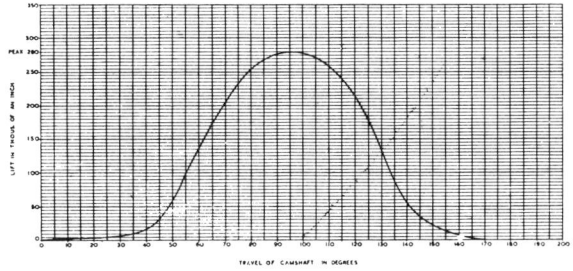

Fig. 13.—Exhaust valve lift

Plate 17.—Oil pump

a die casting in magnesium alloy, machined only in the oil chamber so as to clear the tops of the gear teeth. Oil is directed through drilled holes in the crankcase to the crankshaft, camshaft and push rod bearings, and there are two feeds to the crankshaft centre bearing. Oilways are also drilled in the crankshaft so as to connect the main journals with the big end bearings.

Three holes are drilled round the small end of the connecting rod, lubrication being provided by splash feed.

Oil is also fed under pressure through the push rods to the rocker arm bearings, draining back to the sump through the push rod tubes.

A tubular type oil cooler is incorporated in the system but this can be by-passed to obtain only partial oil cooling. A warning light on the instrument panel indicates when the oil pressure is insufficient.

ENGINE AIR COOLING SYSTEM

Cooling of the engine is arranged by air circulation, and since the engine is fitted at the rear of the vehicle, a blower is provided.

The engine compartment is arranged with a series of slots; these are situated above the rear lid and provide the main air entrance.

The air circulating equipment is particularly noteworthy, being totally enclosed in a metal cowling, air access being arranged at the rear through a 6 in. dia. hole. Air is drawn through this orifice and into the cowling by means of a rotor, which is keyed to the dynamo armature shaft (at the opposite end to the drive), and runs at a speed $1\tfrac{3}{4}$ times that of the engine.

The rotor is a casting of magnesium alloy, and has blades arranged for circulating air by centrifugal action. The oil cooler is enclosed by the cowling and deflectors are arranged for distributing air to each bank of cylinders and the oil cooler. Provision for cooling the dynamo is also incorporated by arranging auxiliary blades on the rotor which draw air through the centre of the dynamo.

The cowling is formed in two halves from steel pressings, lap jointed on the periphery and spot welded together. It is held rigidly by a large magnesium alloy cast flange mounted on the dynamo, and has ducts arranged to enable the air to flow through the dynamo into the atmosphere.

The operation of the air cooling system is extremely noisy.

DYNAMO PULLEY

The pulley is of the split type, formed of two steel pressings with spacing washers between them, so that by varying the number of washers the distance between the pulley flanges can be altered so as to give a certain amount of adjustment to the belt tension.

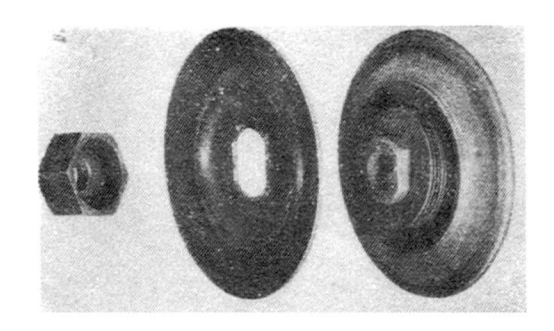

Plate 18.—Dynamo pulley construction

INLET MANIFOLD

The inlet manifold has an extremely small bore and is made from tubing with fixing flanges welded on the ends. A "hot spot" chamber (formed from two half-pressings) is welded at the centre and has a flange welded on top for the carburettor fixing. The hot spot has a small bore pipe welded on the underside, extending at each end into the exhaust pipes.

CARBURETTOR

A single downdraught Solex Type 26FVI carburettor is fitted; this is manufactured by Volkswagen, Germany. It consists of two main die castings, i.e., float chamber and combined float chamber cover and throttle tube; a butterfly throttle is employed. Generally, it resembles the Solex design employed in this country.

The choke is manually operated and an offset butterfly is used. This has an automatic

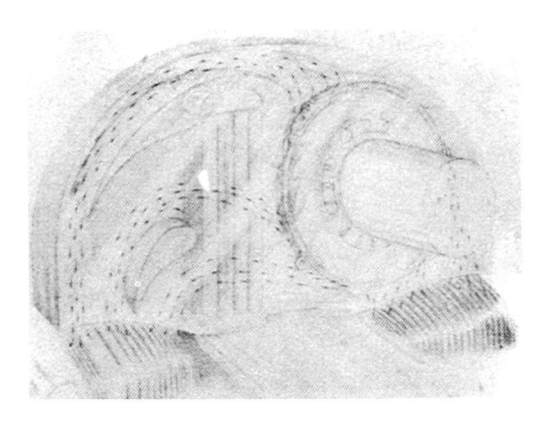

Fig. 14.—Engine air cooling system, phantom view

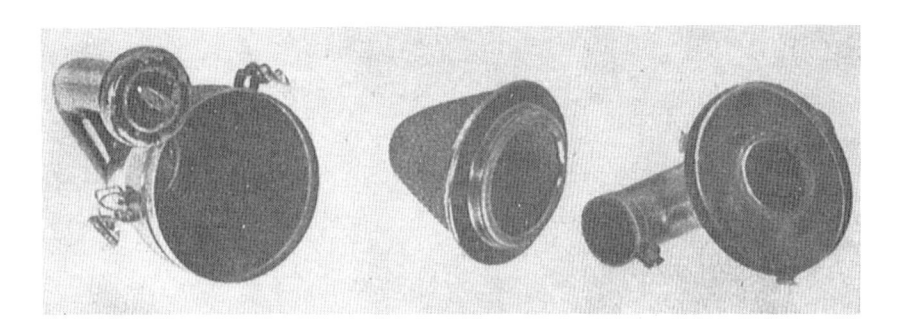

Plate 19.—Air cleaner, exploded view.

spring-loaded valve of Zenith type, which limits the value of depression while starting.

AIR CLEANER

An oil type air cleaner is fitted and is installed alongside the carburettor at about the same height to provide a compact engine unit. The filter is conically shaped, air entering tangentially just above the base of the filter body so as to create a swirling action. Any dust contained in the air is thrown against the wall by centrifugal action and is retained by the oil. The grit gradually settles to the bottom, being removed from time to time. Oil may possibly be drawn with the air into the filter, but as the passages are very fine the restriction makes it practically impossible for any dust to pass through without being caught by the oil. The filter consists of a very fine wire element housed in a wire cage and is removable.

EXHAUST SYSTEM

Four exhaust pipes lead from the cylinder heads to two cylindrically shaped silencers situated in front of the engine at the outer ends. The silencers are supported by the exhaust pipes and the whole system forms part of the engine assembly.

ENGINE UNIT MOUNTING

The engine unit, including the gearbox and axle differential, is suspended on rubber mountings arranged to permit a small rotational movement of the unit about an axis passing through its centre of gravity and coinciding with the centre-line of the vehicle in plan. This is arranged to absorb the torque reactions, but owing to the half-swing axle type of suspension employed, it is not considered good practice. Interaction takes place between the half-axle and the engine due to wheel movement or roll and engine reaction ; consequently the engine movement is restrained to some extent and limits the value of the rubber mounting.

Comparatively small movement is permitted on the engine mounting and this may have been purposely designed, having regard to the interaction. The tension on the rubber mounting is adjustable. The engine mountings consist of two rubbers vulcanised on a steel bracket bolted to the prongs at the rear end of the backbone fork which, in turn, supports the engine around the portion of the crankcase adjacent to the flywheel. This forms the main support for the engine unit on the backbone, one other point of attachment being situated at the forward end (or nose) of the unit. A rubber ring type mounting is fitted, acting principally as a location for the unit. It is housed inside a casting at the centre of the tubular cross member, i.e., centrally in the backbone fork.

ENGINE DATA

Engine No. : 001346.
Make : Volkswagen.
Type : Four cylinder horizontally opposed O.H.V. four-stroke petrol engine.
R.A.C. Rating : 12·2.
Cooling : Air.
Bore : 70 mm. (2·756 ins.).
Stroke : 64 mm. (2·52 ins.).
Total Piston Displacement : 985 c.cm. (60 cu. ins.).
Firing Order : 1, 4, 3, 2.
Direction of Rotation : Anti-clockwise (looking from flywheel end).
Valve timing (·005 in. tappet clearance) : Inlet opens, 14½° before T.D.C. ; Inlet closes, 66° after B.D.C. ; Exhaust opens, 64° before B.D.C. ; Exhaust closes, 11° after T.D.C.
Tappet Clearance : Inlet, ·005 in. ; Exhaust, ·005 in.
Valve Lift (Inlet and Exhaust) : 7 mm. (·275 in.).
Valve Rocker Lever Ratio : 1 : 1.
Throat Dia. (Inlet and Exhaust) : 24·5 mm. (·965 in.).
Angle of Seating (Inlet and Exhaust) : 45°.
Valve Spring Pressure (Inlet and Exhaust) : 44·8 lbs. closed ; 67·2 lbs. open.
Carburettor Type : Solex 26 V.F.I.
Carburettor Choke : 26 T.
Carburettor Jet Sizes : Main 120, Correction 185, Emulsion Tube 100, Pilot 45.
Volume of Combustion Chamber : 58·9 c.cm. (3·59 cu. ins.).
Compression Ratio : 5·2 : 1. This was obtained by measuring the capacity of the combustion chamber. An instruction book from a captured vehicle gave figures for light passenger carrying vehicle model K.1 (Type 82) as 5·8 : 1.
Air Cleaner : Oil Type.

ENGINE SIZES

Crankshaft
Front Steady Journal : dia. 40 mm. (1·57 ins.), length 18 mm. (·709 in.).
Main Front Journal : dia. 50 mm. (1·968 ins.), length 20 mm. (·787 in.).
Main Centre Journal : dia. 50 mm. (1·968 ins.), length 20 mm. (·787 in.).
Main Rear Journal : dia. 50 mm. (1·968 ins.), length, 29 mm. (1·14 ins.).

Connecting Rod
Big End Bearing : dia. 50 mm. (1·968 ins.), length 22 mm. (·866 in.).
Small End Bearing : dia. 20 mm. (·787 in.), length 22 mm. (·866 in.).
Centres : 129 mm. (5·078 ins.).

Pistons
Nominal Dia. : 70 mm. (2·756 ins.).
Overall Length : 71 mm. (2·795 ins.).

Piston Rings
Quantity of Rings per Piston : 2 compression and 1 scraper.
Width : top, 3 mm. (·118 in.) ; centre, 3 mm. (·118 in.) ; scraper, 3·75 mm. (·147 in.).

Gudgeon Pin
Diameter : 20 mm. (·787 in.).
Length : 60 mm. (2·362 ins.).

Flywheel
Diameter : 260 mm. (10·24 ins.).
Number of Teeth : 109.

Starter Pinion
Number of Teeth : 9.

Oil Pump
Gears : pitch dia., 27 mm. (1·06 ins.) ; width of teeth, 17 mm. (·67 in.) ; number of teeth, 9.

Camshaft Drive

	Crankshaft Gear.	*Camshaft Gear.*
Pitch dia. :	63 mm. (2·48 ins.)	127 mm. (5 ins.)
Helix angle :	23° L.H. Spiral	23° R.H. Spiral
Face Width :	20 mm. (·787 in.)	20 mm. (·787 in.)
Number of Teeth :	26	52

Valves
Inlet and Exhaust Stem Dia. : 7 mm. (·275 in.)
Inlet and Exhaust Overall Length : 100 mm. (3·94 ins.).
Inlet and Exhaust Angle of Seat : 45°.
Inlet and Exhaust Head Dia. : 28 mm. (1·108 ins.).

Induction Pipe
Bore (to Carburettor) : 24 mm. (·945 in.).
Bore of Branches : 19 mm. (·748 in.).

ENGINE MATERIAL ANALYSIS

Crankshaft : *Mn.*, 1% (approx.) ; *Cr.*, more than 0·5% ; *Ni.*, Nil ; *Mo.*, 0·2%/0·4% ; *Hardness*, Brinell 302.

Flywheel : *Mn.*, 1% (approx.) ; *Cr.*, more than 0·5% ; *Ni.*, Nil ; *Mo.*, Nil ; *Hardness*, Brinell 228.

Connecting Rod : *Mn.*, 1% (approx.) ; *Cr.*, more than 0·5% ; *Ni.*, Nil ; *Mo.*, Nil ; *Hardness*, Brinell 228.

Valves (Inlet and Exhaust) : *Mn.*, less than 1% ; *Cr.*, more than 3% ; *Ni.*, Nil ; *Mo.*, more than 0·6% ; *Hardness*, Brinell 325 Stem, Rockwell C.63 Tip.

Crankcase : Magnesium Base Alloy.

Piston : Aluminium Alloy ; *Hardness*, Brinell 106.

Cylinder : Grey Cast Iron containing Chrome and Copper ; *Hardness*, Brinell 251.

Cylinder Head : Aluminium Silicon Alloy, Silicon content 10% to 12%.

Camshaft : Cast Iron containing Copper chilled on Cams ; *Hardness*, Brinell 228 (Shaft), Rockwell C.52 (Cam Nose).

Camshaft Gear : Magnesium Base Alloy.

Crankshaft : Impervious to file.

Valve Seat Inserts : Bronze Alloy.

Crankshaft Journals : Lead Bronze.

Connecting Rod Big End Bearings : Babbitt Metal.

Push Rod (Tube) : Aluminium Alloy.

Push Rod (Head) : Hardened Steel ; *Hardness*, Rockwell C.61.

ENGINE WEIGHTS

Crankcase and Studs : 27 lbs.
Cylinders : 3·875 lbs.
Cylinder Head (with Inserts and Guides) : 7·875 lbs.
Crankshaft with Gear : 14 lbs.
Connecting Rods (with Small End Bush) : Total, ·983 lb.
Cap and Bolts : Small End, ·233 lb. ; Big End, ·750 lb.
Piston : ·485 lb.
Piston Rings (one set) : ·07 lb.

Gudgeon Pin : ·172 lb.
Push Rod : ·22 lb.
Rocker Arm (complete Adj. Pin) : ·203 lb.
Valve Springs (Inner) : ·031 lb.
Valve Springs (Outer) : ·063 lb.
Valve : ·094 lb.
Spring Cup and Collar : ·023 lb.
Total Weight of Engine including Blower Equipment, less Clutch Unit : 170 lbs.
Flywheel, Clutch and Driven Disc : 25·5 lbs.

CLUTCH

Make : Fichtel and Sachs-Komet.
Type : K.10.

The design of the clutch is similar to that made by Borg and Beck, the principal difference being on the driven plate which is solid instead of having a spring cush drive as generally used on the Borg and Beck clutch; other variations are mainly constructional. The clutch is of the single plate, dry disc type, no adjustment for wear being provided in the clutch itself. An individual adjustment is provided for initially locating each release lever, for manufacturing purposes only. Three release levers are fitted, and these are interconnected to a centre plate by means of spring hooks. This is operated by a graphite release bearing housed in a neatly designed steel pressing which is connected by means of spring clips to the operating lever. The latter consists of two pressed levers welded to a steel bar, one end only being supported, and having its bearing direct in the gearbox casing. The clutch-driven plate slides on serrations instead of the more usual spline fitting. It was observed that no provision for ventilation of the clutch was arranged, since it was completely enclosed in an aluminium casing directly behind the rear axle differential.

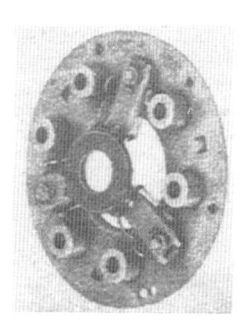

Plate 20.—Clutch unit

CLUTCH DATA

Make : Fichtel and Sachs
Number of Friction Surfaces : 2.
Outer Dia. : 180 mm. (7·09 ins.).
Inner Dia. : 125 mm. (4·92 ins.).
Number of Plates : 1.
Total Area of Friction Surface : 28·5 sq. ins.
Pressure at Release Bearing : 340 lbs.
Clutch Toggle Leverage : 4·4 : 1.
Clutch Control Leverage : 10·44 : 1.
Overall Leverage to Pedal : 44·4 : 1.
Clutch Shaft Serrations : Outside dia. (shaft) 20 mm. (·79 in.) ; Inside dia. (clutch member), 18·5 mm. (·767 in.). Number of serrations, 24.

GEARBOX

General Description

The gearbox forms part of a unit which includes the axle differential and clutch operating mechanism, the whole being housed in a magnesium base alloy casting which is split longitudinally and vertically through

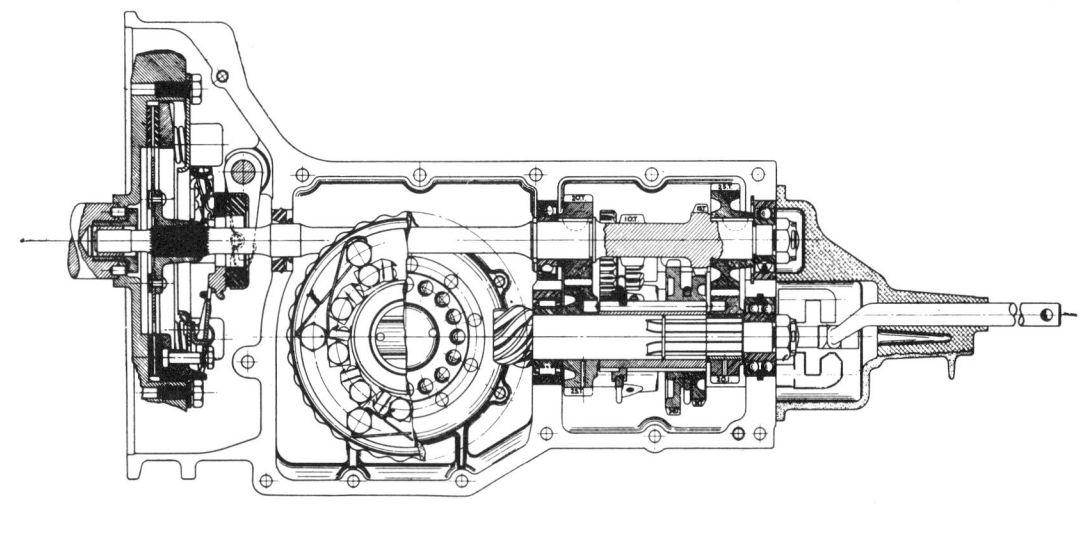

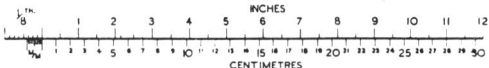

Fig. 15.—Gearbox and clutch (longitudinal section)

the mainshaft and bevel pinion shaft. The gearbox is situated directly in front of the rear axle, being driven from the engine through an extended mainshaft which passes over the axle, and is supported at the crankshaft end in a self-lubricating bronze bush. The clutch is positioned immediately behind the axle and enclosed in the same casting.

The gearbox is of the two-shaft type, having four forward and one reverse speeds. Gear changes are effected through an unusual type of engagement for third and fourth speed, the sliding gear type engagement being used for the first, second, and reverse speeds. Two trains of single helical cut gears are constantly in mesh and provide the third and fourth speeds alternatively; the rest of the gears are straight spur type. Dismantling and assembly were very easy on account of the split casing, which enabled the mainshaft and bevel pinion shaft to be assembled as complete units. The usual type of selector controls are fitted, these being enclosed in a cover in front of the gearbox and operated by a remote control lever of the ball change type.

Gear Engagement.—The gear engagement on the bevel pinion shaft provides a noteworthy feature. Nine pins fit in corresponding semi-circular grooves in the sleeve which is splined on to the shaft, and also in semi-circular grooves in the centre member. The pins act as the driving medium for the centre member, which gives either first or second speeds when in mesh. To obtain third or fourth speeds the pins are moved along the grooves by the selector and the ends engage with corresponding holes in one of the constant mesh gears. This design replaces the normal dog engagement and consequently reduces the overall length of the gearbox; it also provides easy engagement which is simple to produce.

Constructional Details of Gearbox

Gearbox Casing.—The gearbox casing is a magnesium base alloy sand casting formed in two halves dowelled and bolted together. Apart from the gears it also houses the rear axle differential, clutch unit, and carries the starter motor on the R.H. half. One oil filler serves both gearbox and axle.

Condition of Gearbox

Very little signs of wear were observed, and the condition was good.

Mainshaft Assembly

The mainshaft is produced from an electrically upset forged bar having the first and second

Plate 21 (above).—Gearbox and gears (assembled in L.H. interior casing)

Plate 22 (right).—Gearbox, showing selector control and reverse wheel (L.H. interior casing)

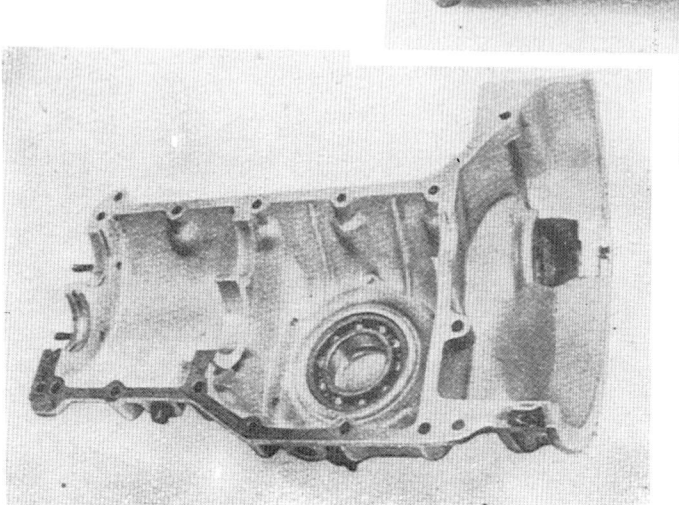

Plate 23 (left).—Gearbox casing, R.H interior

speed teeth integral; the third and fourth speed gears are keyed on, ball races being fitted at each end and secured by a nut and split pin to form a complete assembly.

Bevel Pinion Shaft Assembly

The bevel pinion shaft has a roller bearing at the pinion end and a double row ball race at the opposite end. The third and fourth speed idler gears run directly on the shaft, interposed between them being a composite spline fitting on which slides the combined first and second speed gear. These are secured by nut and split pin to form an assembly.

Gear Selector Control

Three gear selector shafts are housed in the left-hand half of the gearbox casing, the operating lever being situated at the front end in a magnesium base alloy die cast cover which is bolted to the gearbox casing. The nose of the cover is fitted with a rubber block, thus providing a mounting which locates the complete power unit in the backbone chassis.

Gear Control

A ball change type of remote gear control lever is used, pivoting in a pressed steel housing bolted on top of the backbone. A tubular control rod connects the gear lever with the selector, the control being located inside the backbone. The front end of the rod is supported inside the backbone and the usual type of ball and cup fitting between the lever and rod is employed.

GEARBOX DATA

GEAR SIZES

Gear No.	Gear Ratio	Pitch Dia. (approx.) Driving	Pitch Dia. (approx.) Driven	Gear Width Driving	Gear Width Driven	No. of Teeth Driving	No. of Teeth Driven	Helix Angle
1st	3·6 : 1	28 mm. (1·10″)	96·5 mm. (3·80″)	15·5 mm. (·61″)	10 mm. (·787″)	10	36	Straight
2nd	2·07 : 1	41 mm. (1·614″)	83·5 mm. (3·287″)	10 mm. (·394″)	10 mm. (·787″)	15	31	,,
3rd	1·25 : 1	55·5 mm. (2·185″)	69·5 mm. (2·736″)	20 mm. (·787″)	20 mm. (·787″)	20	25	28°
4th	0·8 : 1	69·5 mm. (2·736″)	55·5 mm. (2·185″)	20 mm. (·787″)	20 mm. (·787″)	25	20	28°
Reverse	6·6 : 1							

BEARING SIZES

	Outside Dia.	Inside Dia.	Width
Mainshaft Ball Race	52 mm. (2·047″)	25 mm. (·98″)	15 mm. (·59″)
,, ,, ,,	,,	20 mm. (·787″)	,,
Bevel Pinion Shaft Roller Race	62 mm. (2·44″)	30 mm. (1·18″)	20 mm. (·787″)
Bevel Pinion Shaft Double Row Ball Race	52 mm. (2·047″)	20 mm. (·787″)	22 mm. (·866″)

GEARBOX MATERIAL ANALYSIS

Part	Mn.	Cr.	Ni.	Mo.	Hardness
Mainshaft	more than 0·6%	more than 0·5%	Nil	less than 0·5%	Rockwell C.62
Bevel Pinion Shaft	,, ,, 0·6%	,, ,, 0·5%	Nil	,, ,, 1·0%	C.60
3rd and 4th Gears	,, ,, 0·6%	,, ,, 0·5%	Nil	Trace	C.62
1st and 2nd Gears	,, ,, 0·6%	,, ,, 0·5%	Nil	less than 0·5%	C.64
Sliding Collar	,, ,, 0·6%	,, ,, 0·5%	Nil	,, ,, 0·5%	C.58
Mainshaft Sleeve	,, ,, 0·6%	,, ,, 0·5%	Nil	Trace	C.62
Selector Shaft	,, ,, 0·6%	,, ,, 0·5%	Nil	0·3% approx.	C.64
Selector Fork	,, ,, 0·6%	,, ,, 0·5%	Nil	0·5%	Superficially hard

REAR AXLE DRIVE

General Description

A spiral bevel pinion formed on the gearbox output shaft drives the crown wheel, and this transmits the drive to the road wheels through a " limited slip " type differential which is a noteworthy feature. The drive to each wheel is arranged through half-axle shafts enclosed in tubular casings which act as swinging half-axles. On the outer end a final gear reduction box is fitted and the whole assembly (with the road wheel) oscillates about a " pot " type universal joint, which is housed inside the differential cam rings, and in effect forms the " pot." The axle shafts have forged ends which are hardened and ground and these engage with cylindrical segment members (also hardened and ground), which are located in the " pot " and form the universal driving joint.

Plate 25.—Rear axle reduction gear casing and brake back plate

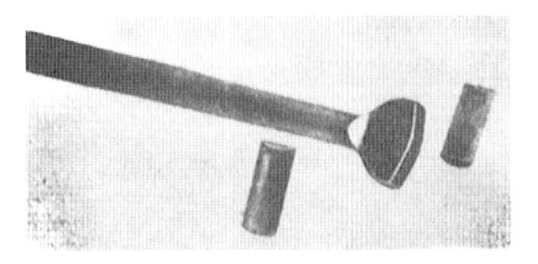

Plate 24.—Rear axle shaft end and cylindrical segment for universal joint

Splines are arranged at the outer end of the shaft for a gear reduction drive. This consists of a pair of straight-toothed spur gears completely enclosed in cast iron casings and supported by ball races, filler plugs being provided in each casing for lubrication purposes. The final gear reduction appears to have been introduced primarily in order to obtain lower gear ratios in a simple manner for military vehicles. Furthermore, it provides an increased ground clearance for the axle casing and power unit, which is an advantage when travelling across country. The differential unit is mounted on ball races which are housed directly in the axle casing and it was observed that the two half-axle shafts not only formed component parts of the differential unit, but they cannot be removed without taking the axle and gearbox unit out of the chassis and dismantling it. The axle casing consists of steel tubing, to which is welded a spherical end with both internal and external polished surfaces. This is held between the axle casing and a magnesium base alloy cover and forms the location about which the axle oscillates. A concertina form of rubber cover is fitted to prevent ingress of dirt to the seating.

Detail Description

Differential. — The differential gear is a " limited slip " type, being a cam form which follows the same principle as the one developed by a German gear manufacturer, Zahnradfabrik Friedrichshafen, in 1931. This is patented by Gottfried Weidmann, German Patent Spec. No. 639876 and British Patent Spec. No. 431020. This type of differential offers considerable advantages when the vehicle is travelling across country.

Plate 26.—Rear axle tube and shaft complete

Plate 27.—Differential and axle shaft assembly

The phantom view, Fig. 16, shows the construction of the differential mechanism. This consists of two hardened steel disc members or cam rings which are mounted fast on to the axle shafts and are provided with recessed cam surfaces. Between these surfaces are interposed 17 hardened steel dog transmitting elements or plungers which are supported in holes or seatings in a hardened steel carrier member. The plungers are axially movable into the recesses, to permit differential action of the cam rings.

The above unit is held in a steel casing formed in two halves and secured to the plunger carrier and crown wheel by bolts and nuts, pressed steel scoops being attached for lubrication purposes. Thrust washers of plastic material are interposed between the cam rings and the differential casings.

The differential behaves as follows under various conditions :

(a) When one of the wheels meets so high a resistance as to exert a strong braking force—such as would occur when the vehicle deviates from a straight course—then the cam ring driving the inner road wheel may come practically to a standstill and the locked cams of this ring strongly resist the push of the sliding plunger. At the same time the cam ring connected to the outer road wheel offers less resistance and the ends of the plungers abutting against these cams push them forward, simultaneously moving up and down over the cams of the ring which has the higher resistance in the inner wheel. Owing to the friction of the sliding plungers

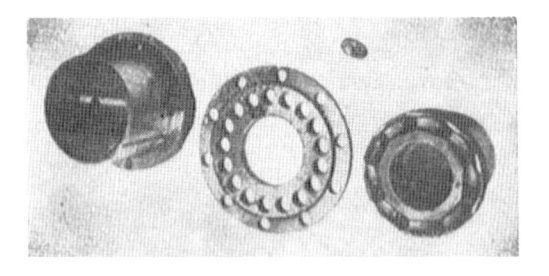

Plate 28.—Differential, exploded view

in their carrier and on the cams, this action takes place only when a relatively powerful resistance is met by one of the road wheels. Minor resistances leave the drive unaffected as the road wheels operate as though they were attached to a solid axle.

(b) When each rear wheel of the vehicle offers the same resistance, the plungers rotating with the carrier driven by the engine will take both cam rings with it so that both wheels rotate at the same speed.

(c) Owing to the driving wheels being relatively rigid one with the other, starting on slippery ground is facilitated.

(d) When both road wheels are raised from the ground and the crown wheel is prevented from rotating, if one wheel is revolved the other wheel turns in the opposite direction, functioning in a similar manner to the usual type of differential. It differs only in that in the conventional type of differential, wheel speeds are identical, whereas in this

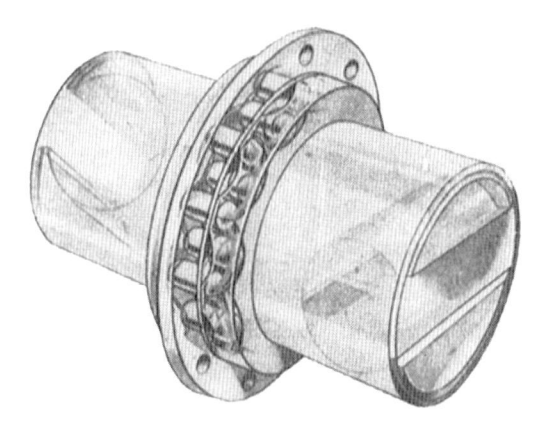

Fig. 16.—Differential, phantom view

case one wheel revolves faster than the other in the ratio of 9 turns to 8 turns. Therefore the torque transmitted to one cam ring and its corresponding half-axle exceeds that transmitted to the other in the ratio of 9 : 8, i.e., $12\frac{1}{2}\%$.

Referring to the development drawing, Fig. 17, it will be observed that ring A has eight and ring B has nine concave surfaces spaced equidistantly ; this variation enables several of the surfaces to be located simultaneously in a position to be driven by the corresponding plungers as the drive picks up. This is known as the " locked " position. If the cam rings were provided with the same number of inclines, this would give all the plungers identical positions relative to all inclines and lead to floating (or slip) if the plungers adapted to drive one of the cam rings were located at the apices of the inclines.

The operation of the differential is substantially as follows : Rotary movement applied to the bevel wheel is conveyed to the carrier and its plungers, which, in turn, drive the two cam rings in such a way as to permit differential action.

Assuming that one of the cam rings is stopped, the plungers continue to drive, during their rotary movement, the other cam ring. This is caused to rotate at the same speed as the bevel wheel plus a further increase in speed due to the arcuate movement of the ends of the plungers on the inclines of the stationary cams.

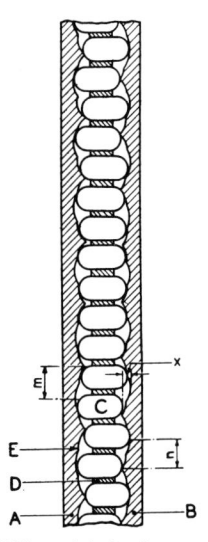

Fig. 17.—Differential, development drawing

At this instant, in addition to their rotary movement, the plungers move transversely in their seatings in the carrier. This particular movement of the plungers operates through reaction, by bearing against the stationary cam ring and imparting to the moving cam wheel an additional angular velocity so that the final speed of the moving cam wheel is substantially twice that of the bevel wheel. By way of example, assuming that cam ring A is stationary, at each movement of the carrier D equal to the circumferential length m, the plunger which, at the beginning of the movement, is at the bottom of the incline E, will, at the end of the movement, be at the succeeding apex. The plunger has therefore been subjected to a movement towards the cam wheel B in such a manner that this wheel has been caused to move relative to the plunger through a distance n, corresponding to one half of its inclines. The resultant movement of cam wheels B to A is finally m plus n.

If, now, instead of being stationary, one of the cam rings has a movement relative to the other (which is the condition when the vehicle travels round a curve), the carrier D which is assumed to rotate at constant speed, imparts to the two cam rings a difference in speed proportional to the difference of the paths travelled by the road wheels.

The plungers C always engage with the cam ring of which the speed is reduced, so as to transmit to the other cam ring an increased speed of $\frac{m}{n}$ or $\frac{n}{m}$ of the slowing down of the other.

When the plungers, by engaging with the cam wheel which rotates at the slower speed, are adapted to transmit to the other cam wheel a supplementary speed, at this moment it must supply the bevel wheel with supplementary work for overcoming the relatively high friction of the plungers which bear against the cam wheel moving at the lower speed, and consequently always tends to drive this cam wheel at the same speed as the bevel wheel. Thus the differential movement is retarded.

Condition of Differential

The holes (or plunger seatings) in the carrier showed signs of considerable wear — i.e., scuffing—and the plungers were a slack fit in the seatings. Considerable hammering had occurred on the spherical ends of the plungers, although the cams were in good condition—this being the case generally. During road tests with the vehicle the differential was observed to be noisy.

Rear Hubs

The rear hubs are in one piece with the brake drums, being made of malleable cast iron. The road wheels are attached to the side of the brake drums in the same manner as those at the front, the only major difference being the spline fitting required on the rear hubs for fixing to the axle shaft.

REAR AXLE DRIVE DATA

Spiral Bevel Drive:

Part.	Pitch Dia.	No. of Teeth.	Spiral Angle.	Face Width.
Pinion …	44 mm. (1·73″)	7	55°	22 mm. (·87″)
Crownwheel	165 mm. (6·50″)	31	55°	22 mm. (·87″)

Hub Spur Gear Drive:

Driving Pinion	64 mm. (2·52″)	15	Straight	30 mm. (1·18″)
Driven Wheel	87 mm. (3·43″)	21	Straight	30 mm. (1·18″)

Gear Ratios:

Bevel gear, 4·43 : 1; spur gear, 1·40 : 1; overall axle gear ratio, 6·20 : 1

Rear Axle Bearing Sizes:

Part.	Outside Dia.	Inside Dia.	Width.
Differential Case R.H. Ball Race …	90 mm. (3·54″)	50 mm. (1·97″)	20 mm. (·79″)
Differential Case L.H. Ball Race …	90 mm. (3·54″)	50 mm. (1·97″)	11 mm. (·43″)
Hub Driving Pinion Ball Bearing (Inner)	72 mm. (2·83″)	30 mm. (1·22″)	19 mm. (·75″)
Hub Driving Pinion Ball Bearing (Outer)	62 mm. (2·40″)	25 mm. (·98″)	17 mm. (·67″)
Driven Gear Ball Bearing (Inner) …	72 mm. (2·80″)	30 mm. (1·22″)	19 mm. (·75″)

Angular Movement of Half Axle:

Laden position $9\frac{1}{4}°$

At laden position 0°; at full bump $6\frac{1}{2}°$; at full rebound $9\frac{1}{4}°$; total angular movement $15\frac{3}{4}°$.

Rear Axle Weights:

Transmission Unit with Clutch and Starter less Brake Shoes and Drums : 147 lbs.

REAR AXLE MATERIAL ANALYSIS

Part.	Mn.	Cr.	Ni.	Mo.	Hardness.
Crown Wheel …	0·6/1·0%	more than 0·5%	Nil	Trace	Rockwell C.62
Differential Casing	0·6% approx.	Nil	Nil	less than 0·5%	
Differential Cam Ring	0·6% approx.	0·5%	Nil	less than 0·5%	("Pot") (fitting Brinell 338) (Outside dia. Rockwell C.58)
Differential Cam Ring Sleeve					
Differential Plunger Carrier	0·6% approx.	more than 0·5%	Nil	less than 0·5%	Rockwell C.60
Differential Plunger					Rockwell C.64
"Pot" Joint Cylindrical Segment	less than 1·0%	Nil	Nil	less than 0·5%	(VPN.50 kilos. 878)
Axle Shaft … …					(Forged end VPN.50 kilos 610)
	less than 1·0%	1·0% approx.	Nil	less than 0·5%	Splined end Brinell 293
Hub Driving Pinion	0·6%/1·0%	less than 1·0%	Nil	Trace	Rockwell C.60
Hub Driven Gear…	1·0%	less than 1·0%	Nil	,,	Rockwell C.60
Hub Driven Gear Shaft	0·6%/1·0%	less than 1·0%	Nil	,,	Brinell 281
Hub Gear Casing …	0·6%/1·0%	Nil	Nil	,,	Brinell 148
Axle Shaft Tubular Casing					Brinell 229

Fig. 18.—Rear axle and gearbox (cross section)

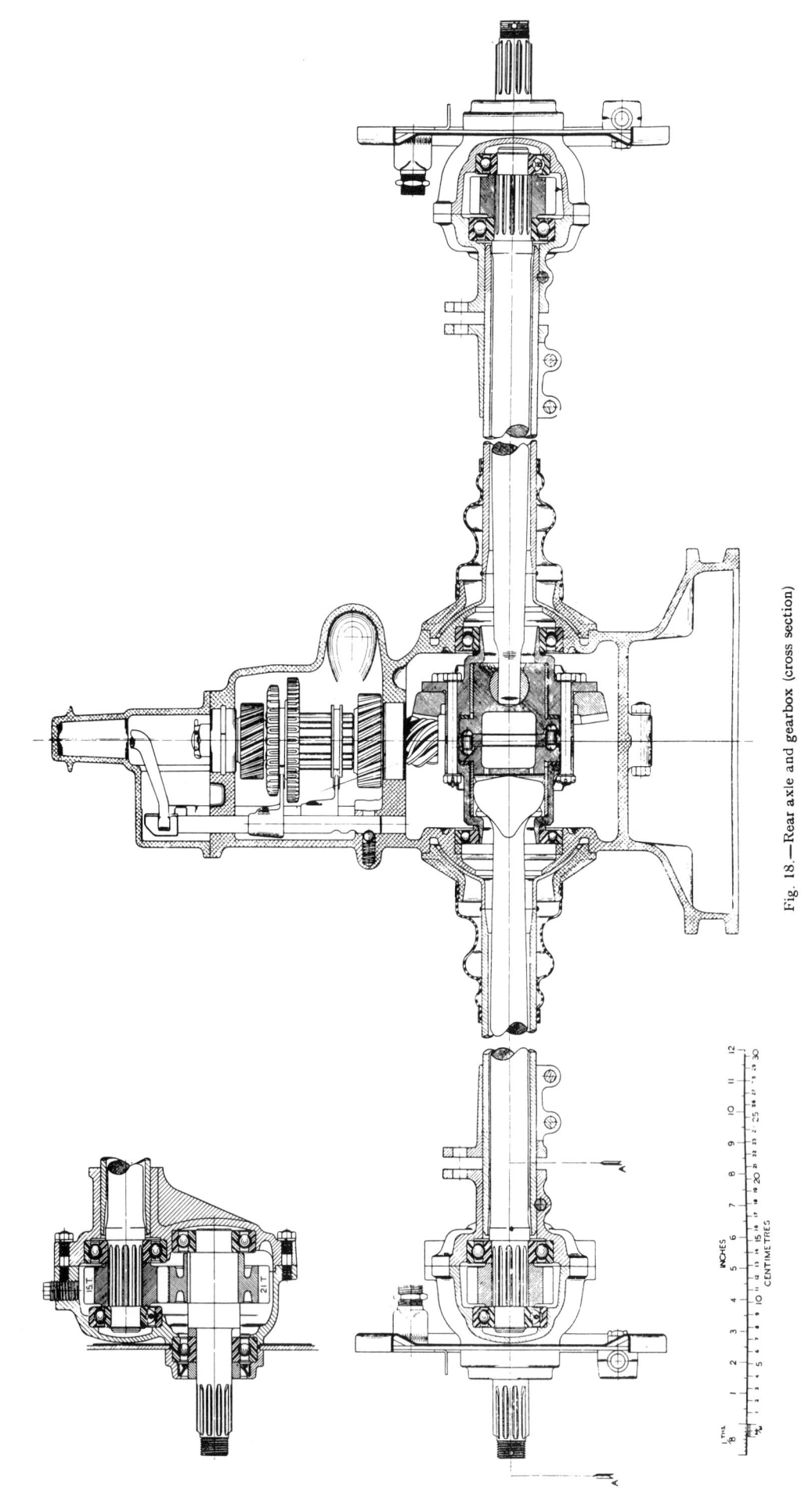

REAR SUSPENSION

General Description

Independent rear wheel suspension is fitted. It consists of torsion bars situated transversely in front of the rear axle and enclosed in a tubular cross member forming part of the underframe. Longitudinally swinging arms are attached to the outer ends, which have a trailing action. In turn, these carry the outer ends of the swing half-axle casings to which they are rigidly connected. The suspension linkage for each wheel thus represents two sides of a right-angled triangle, the wheel being situated at the right-angled corner. Consequently the wheels oscillate about an axis formed by the hypotenuse of a triangle; the hypotenuse intersects the centres of the swing axle spherical seating and the longitudinal arm bush. This structure is suspended on the torsion bars and owing to these being arranged transversely, the longitudinal arms are made flexible to allow for misalignment. This form of suspension is notable for its simplicity; it entails only one universal driving joint per half-axle and the longitudinal arm which serves the dual purpose of the torsion bar suspension arm and the brake torque arm. Furthermore, only two bearings are used in the linkage for each wheel, i.e., the half-axle spherical seating and the torsion bar rubber bush. Torsional movement is restricted by stops arranged on brackets welded on to the tubular cross member, which also carry double acting type shock absorbers, of the detachable hydraulic lever arm type. No means of adjustment is provided for the torsion bars, winding of which may occur on account of initial loading and fatigue, except by dismantling the torsion bars and re-positioning on the serrated fitting. Comparatively short torsion bars of robust circular section are used and consequently the suspension is fairly stiff—much stiffer than that fitted to the front wheels.

Means are provided whereby the torsion bars can be extracted in the event of failure, and this without disturbing the road wheel and longitudinal arm. The outer ends of the torsion bar are supported in rubber bushes which are fairly hard and conically seated, so that they are wedged on both the inside and outside diameters.

Constructional Details

Two torsion bars of circular section are fitted, and these have serrations on each end, those at the inner end being smaller in diameter than at the outer, facilitating assembly and dismantling without disturbing the road wheel and longitudinal arm.

The torsion bars are enclosed in a substantial tubular cross member situated transversely and forming part of the underframe structure. At the centre of this member a malleable iron casting is welded in position and this has a serrated hole for the torsion bars. Malleable cast iron brackets are welded to the extremities of the tubular members, serving the purpose of housing rubber bushes for supporting the torsion bars, restricting the wheel movement by means of stops acting against the longitudinal arms and providing an anchorage for the shock absorbers. The longitudinal arms are of built-up construction, each consisting of a pressed steel strip butt welded to a steel stamped boss at one end, the latter being serrated to fit the torsion bar. The method of welding across the strip is shown in the drawing of the rear suspension, being arranged cheaply and efficiently.

The opposite end is fork shaped to fit over the axle casing, facilitates replacements of swing arms failures, and is rigidly secured to the axle brackets by means of three bolts. Swing arms are thus easily replaced in case of failure.

At the extremities of the tubular cross member are fitted magnesium base alloy die cast covers which form the closure for the torsion bars and also support the outer rubber bushes.

The torsion bar swing arm or strut is located above the wheel centre, and it is claimed in

Plate 29.—Rear suspension, R.H. wheel

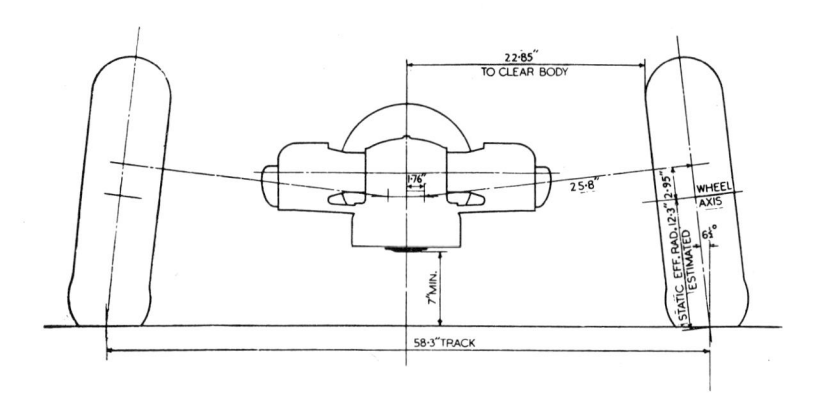

Fig. 19.—Rear suspension, minimum ground clearance

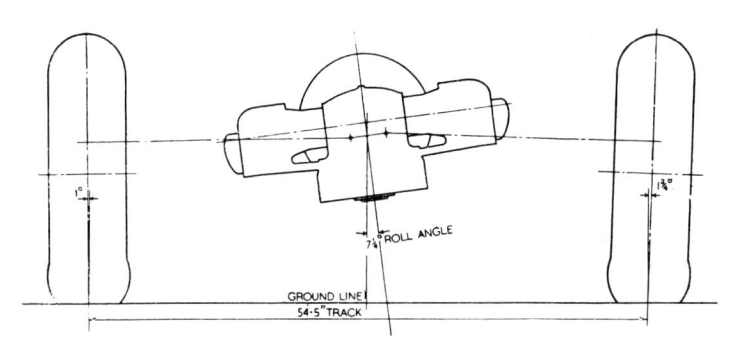

Fig. 20.—Rear suspension, maximum roll position

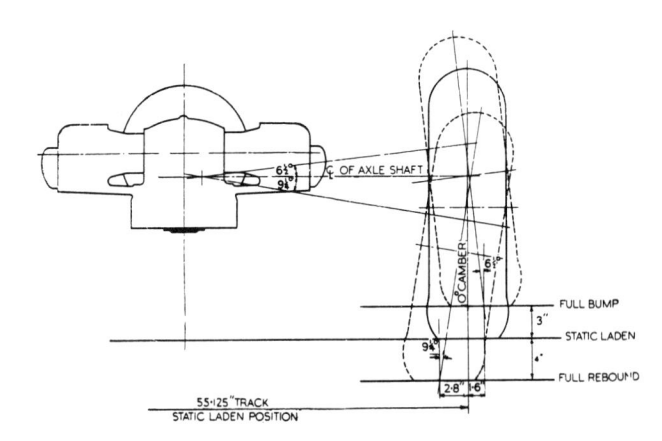

Fig. 21.—Rear suspension, wheel movement

British Patent Spec. No. 544748 taken out by F. Porsche that the stresses in the strut are minimised and consequently a light and cheap construction is possible.

It also appears, however, that improved ground clearance is gained due to the raised position of the torsion bar relative to the vehicle, which is thus more suitable for cross-country work. With the exception of the hub reduction gear, this arrangement would permit all the components to be interchangeable with the original design, as far as can be ascertained.

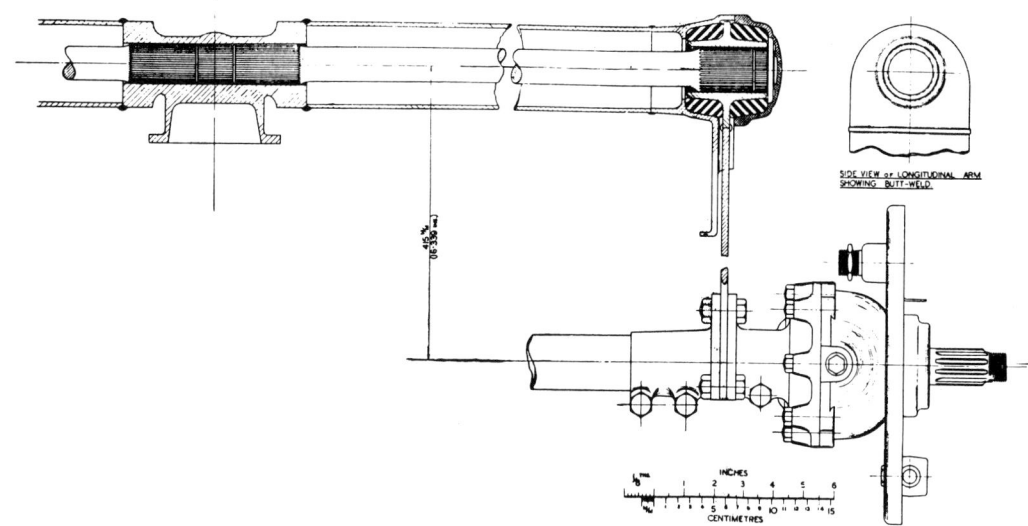

Fig. 22.—Rear suspension, R.H. side, plan

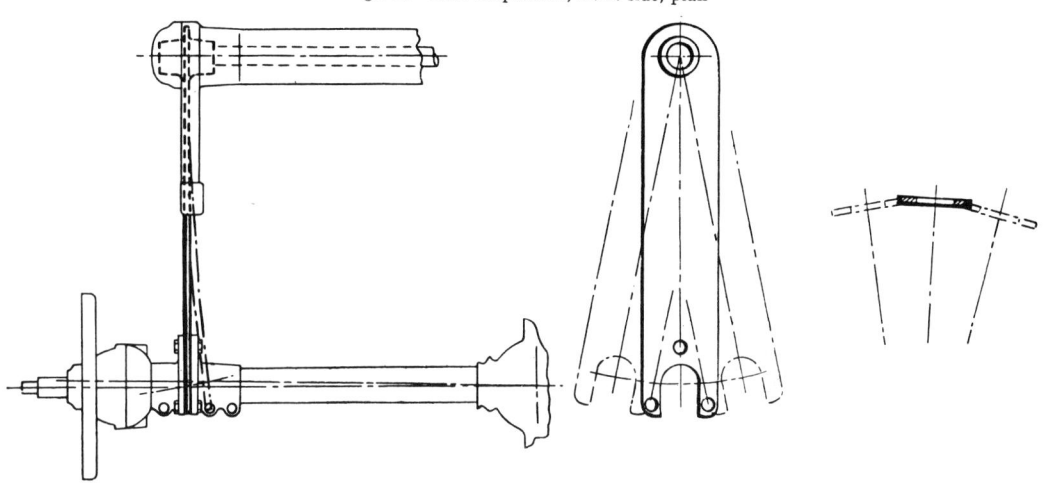

Fig. 23.—Rear suspension, longitudinal link, action diagram

REAR SUSPENSION DATA

Weights.		*Unsprung Weight.*
Rear Axle Shaft	7·34 lbs.	4 lbs.
Rear Axle Casing complete	28·125 lbs.	24·25 lbs.
Rear Hub and Brake Assy.	18 lbs.	18 lbs.
Swing Arm	—	1·72 lbs.
Wheel	12·75 lbs.	12·75 lbs.
Tyre	20·5 lbs.	20·5 lbs.
Inner Cover	3·25 lbs.	3·25 lbs.

Total 84·47 lbs. per wheel

168·94 lbs. both wheels

(169)

Laden Sprung Weight at Rear Wheels

1,598 lbs. − 169 lbs. = 1,429 lbs. = 715 lbs. per wheel.

Swing Arm centres 16·34 ins.

Torsion Bar

Round Bar : 1·093 ins. dia.
Effective Length : 18 ins.
Rate of Spring : 352 lbs. per in./per wheel.
Torsional Stress : 20·2 tons/sq. in.

Wheel Movement

Total : 7 ins.
Static laden to full bump : 3 ins.
Static laden to full rebound : 4 ins.

Track

55·125 ins. (4 ft. 7⅛ ins.) at static laden position.

Wheel Camber

0° at static laden position.

Ground Clearance

With 12·3 effective radius tyre :

Lowest sprung point is the engine sump drain plug.

Ground clearance = 10 ins. static laden position.

FRONT SUSPENSION AND STEERING HEAD
(including Speedometer Drive)

General Description

Independent front wheel suspension is fitted, of a type which can be described as the "Longitudinal Link." It has a trailing action, and is suspended by an ingenious design of torsion bar consisting of four rectangular shaped strips placed together (in lieu of the usual round bar section) a considerable increase in effective length being gained, and in consequence a greater degree of resiliency. Although roughly 170 per cent. less efficient than round bar of equal weight, the arrangement is nevertheless very compact and cheap. There are two torsion bars housed in superimposed cross tubes attached to the underframe; each bar is fixed to the tube at the centre, the links attached to the outer ends carrying the steering head. This forms a parallel link motion imparting purely torsional movement to the bars, which are limited on bump and rebound by stops arranged on the cross tube bracket. Detachable shock absorbers of the hydraulic single-acting barrel type are fitted. No means of adjustment is provided for the torsion bars, winding of which may occur, due to initial loading or fatigue; the fact that the bars are not highly stressed may be the reason for omitting the adjustment.

Although it is understood that Dr. F. Porsche was partly responsible for the design of this vehicle, the front suspension (which follows the same principles as the "Porsche" torsion bar system) differs in regard to the torsion bar, also the connection between the swing arms and the steering head. Instead of a ball seating, the normal "swivel pin" type is used, and this is noteworthy in that the swivel pin is constructed in two parts, thereby enabling the spacing between the swivel-pin centres and also the swing arm crosshead bushes to be as wide as that which would be obtained if ball seatings had been employed.

CONSTRUCTIONAL DETAILS
Front Suspension

The front suspension, including wheels, shock absorbers, steering gearbox unit and track rods, can be assembled as a complete unit and attached to the backbone of the underframe by means of four bolts. The main structure which houses the torsion bars consists of two parallel spaced cross tubes bridged by four pressed steel brackets welded on equidistantly. The two central brackets act as support mountings, whilst the other two at the extremities carry stops for restricting the up and down motion of the wheel, an anchorage is also provided for the shock absorber. The torsion bars, which are in-

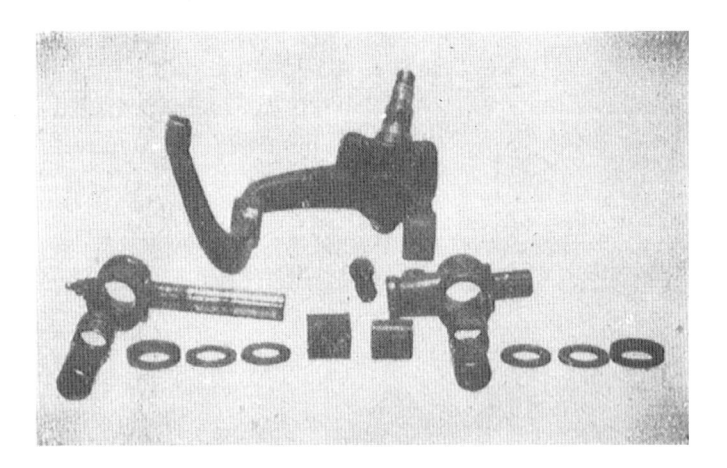

Plate 31.—Steering head, R.H. exploded view

terchangeable, are fixed to collars midway in the cross tubes which, in turn, are held against rotation by indentations in the tube and centre pins (see Fig. 27). At the outer ends the suspension links (which are steel stampings) are secured to the torsion bars by centre pins and square hole fittings. Each link oscillates on two bushes made from a plastic base material; these are supported in the cross tube and facilities for lubrication are also provided by grease nipples screwed into the tubes. Rubber seals prevent the ingress of dirt to these bearings.

Plate 30.—Front suspension unit.

35

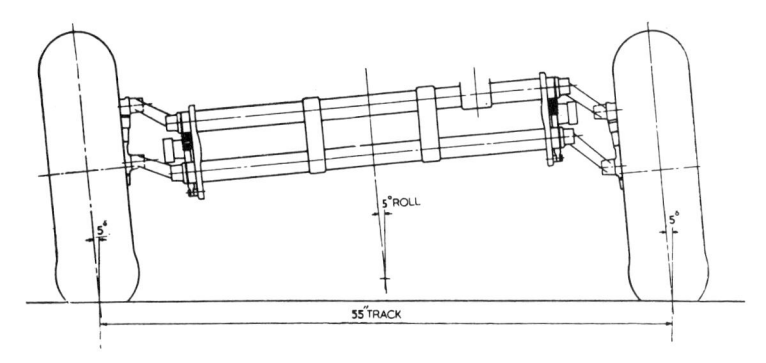

Fig. 24.—Front suspension, maximum roll position

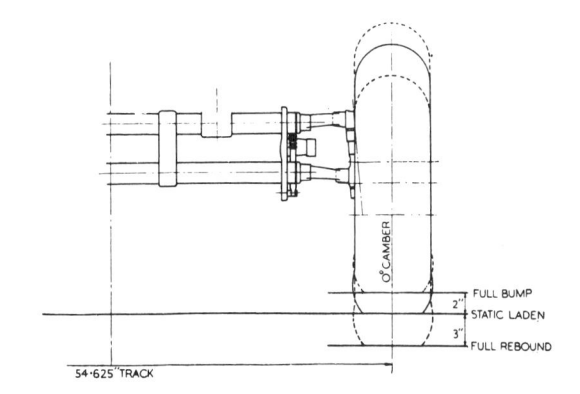

Fig. 25.—Front suspension, wheel movement

CONSTRUCTIONAL DETAILS

Steering Head

The steering head incorporates a swivel pin of built-up construction comprising two parts, each made from a steel stamping, and held together —with the stub axle in position—by means of locating bolts (see Plate 31 and Fig. 26). Each half swivel pin has a cross-head in which plastic base bushes and loose thrust washers are fitted. The complete swivel pin assembly oscillates on these bearings, which are situated transversely on overhung hardened steel pins attached to the swing arms. The pins are adjustable for wear on the thrust faces, employing the same method as is used on the steering gearbox sleeve, i.e., a slow helix groove provided on the outside diameter of the pin, which fits into the swing arm and is located by a pinch bolt. When the pin is rotated it is compelled to move endwise by the pinch bolt, thus taking up end play. The swivel pin bearings are of bronze and the thrust washers on these are of plastic base material. Lubrication is arranged by means of one grease nipple on top of the upper half swivel pin and supplying cross-head bearings, swivel pin bearings and thrust washers. This necessitates fitting the cross heads with steel sleeves having annular grooves, in order to provide oilways for the swivel bearings.

Stub Axles

The stub axles are steel stampings resembling the shape employed for "Reverse Elliot" type axles. They are compact, inasmuch as the steering arms are formed integral.

Front Hubs

The front hubs are of malleable iron cast in one piece with the brake drums. A noteworthy point is the provision of five tapped holes in the back of the brake drum for fixing the road wheels. Fixing is effected by means of set bolts having spherical seatings under the head; no spigot location is arranged. Cup and cone type bearings are used, loose cups and cones being fitted with caged balls. No facilities are provided for greasing the hub bearings. Hub retainer caps are fitted into recesses in the end of the hubs, being held by means of a press fit.

Speedometer Drive

The speedometer drive is neatly housed inside the left-hand stub axle. It consists of two spindles arranged approximately at right angles to one another, having skew gear form of drive. One spindle is disposed along the axis of the stub axle and is driven off the hub cap by means of a split pin, which locates in a slot in the end of the spindle. The other spindle is arranged vertically and close to the swivel pin, so that the position of the speedometer drive cable is not unduly altered by the road wheel steering movement.

Fig. 26.—Front suspension, L.H. side, front view

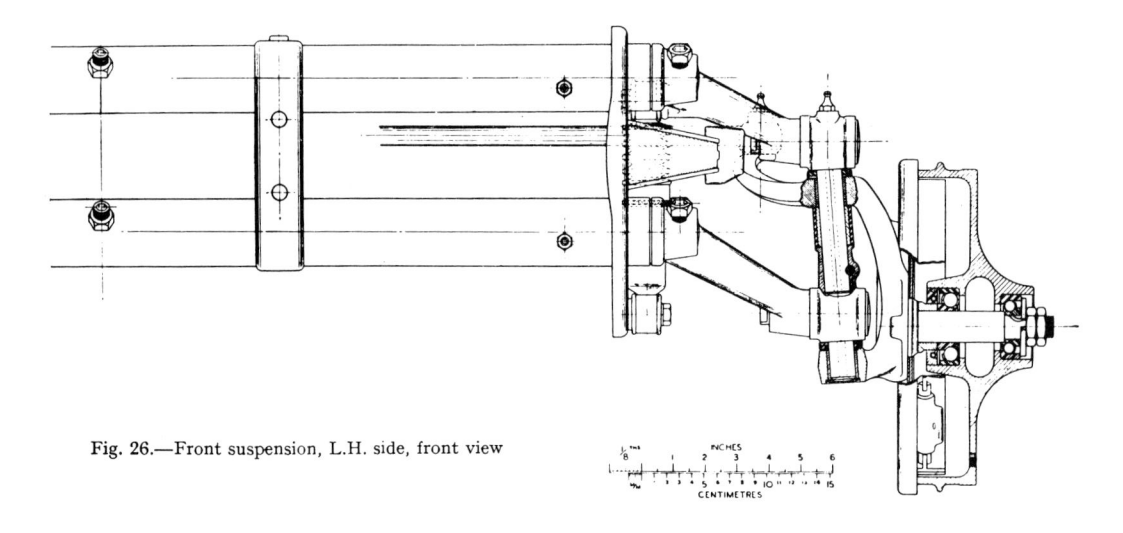

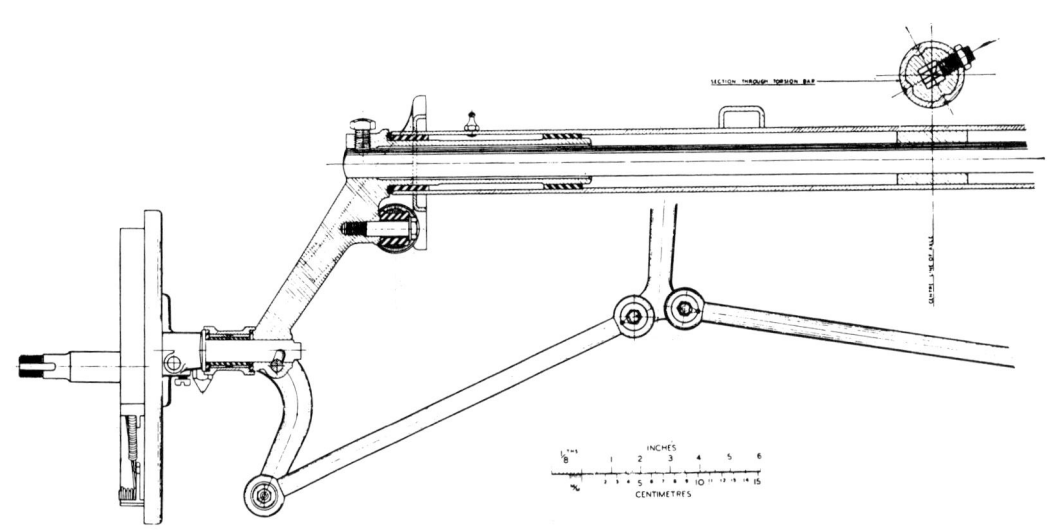

Fig. 27.—Front suspension, L.H. side, plan view

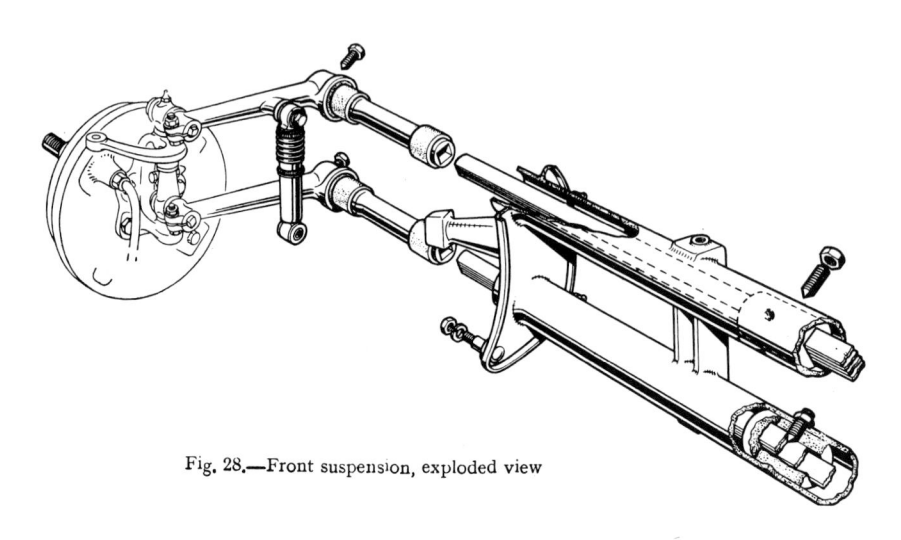

Fig. 28.—Front suspension, exploded view

FRONT SUSPENSION DATA

Weights. *Unsprung Weight.*

Front hub and stub assy.	24·94 lbs. =	24·94 lbs.
Top link	4·44 lbs. =	1·69 lbs.
Bottom link	4·28 lbs. =	1·61 lbs.
Track Rod (short)	1·44 lbs.	
Track Rod (long)	2·78 lbs. =	1·06 lbs. total
Wheel	12·75 lbs.	12·75 lbs.
Tyre	20·5 lbs.	20·5 lbs.
Inner Cover	3·25 lbs.	3·25 lbs.
	Total	65·8 lbs. per wheel
	=	131·6 lbs. both wheels
		(132)

Laden Sprung Weight at Front Wheels

992 lbs. − 132 lbs. = 860 lbs. = 430 lbs. per wheel.

Suspension Links

Centres: 5·875 ins.

Torsion Bar

Four blades comprise one torsion bar; 2 torsion bars used.

Overall effective length: 37 ins. (18½ ins. centreline of vehicle to link).

Width of blade: ·734 in.

Depth of blade: ·183 in.

Rate of spring: Total 130 lbs. per in. per wheel

Swivel Pin Inclination: 4° 30′.

Wheel Camber: 0°.

Castor Angle: 2° 30′.

Tyre Offset (at 12·3 effective radius): 2·31 ins.

Wheel Movement (total): 5 ins.

Wheel Movement (static laden to full bump): 2 ins.

Wheel Movement (static laden to full rebound): 3 ins.

Track: 54·625 ins. (4 ft. 6⅝ ins.).

Swivel Pin Bearings

Diameter: 20 mm. (·79 in.).

Length: 22 mm. (·87 in.).

Bearing Spacing: 5·2 ins. centres.

Suspension Link Bearings

Small: dia., 18 mm. (·71 in.); length, 33·5 mm. (1·32 ins.).

Large: dia., 37 mm. (1·45 ins.); length, 32 mm. (1·26 ins.).

SUSPENSION DATA

Part.	Mn.	Cr.	Ni.	Mo.	Hardness.
Front Suspension					
Torsion Bar	More than 1·0%	More than 1·0%	Nil	Trace	Brinell 415
Rear Suspension					
Torsion Bar	do.	do.	do.	More than 1·0%	do.

SHOCK ABSORBERS

FRONT SHOCK ABSORBERS

The front shock absorbers are fitted between the suspension top link and the chassis cross tube outer bracket, one per wheel being employed. The anchorage consists of rubber bushes having steel inner sleeves.

Construction

The general construction is in the class known as direct acting or telescopic type shock absorbers. Refer to sectional arrangement drawing Fig. 30

It consists essentially of a steel cylinder closed at the bottom, a piston reciprocable in the cylinder with a light loaded non-return valve opening on the up stroke of the piston, and a more heavily loaded valve opening on the down stroke of the piston.

Plate 32.—Front shock absorber, exterior

Attached to the upper end of the cylinder is a deformable rubber bellows having free communication with the cylinder and serving for the accommodation of working liquid at approximately atmospheric pressure.

The piston is an aluminium casting and houses the only two valves in the construction. The main resistance relief valve is a simple steel cone but is spring loaded in an unusual manner as shown on the sectional arrangement. The valve spring (a coil spring) is not concentric with the valve and only the outside of the bottom coil presses on the valve. Presumably this is to get a larger diameter spring with a low rate, and so limit the build up in resistance with an increase in speed to as low a value as possible. The low resistance recuperating or transfer valve,

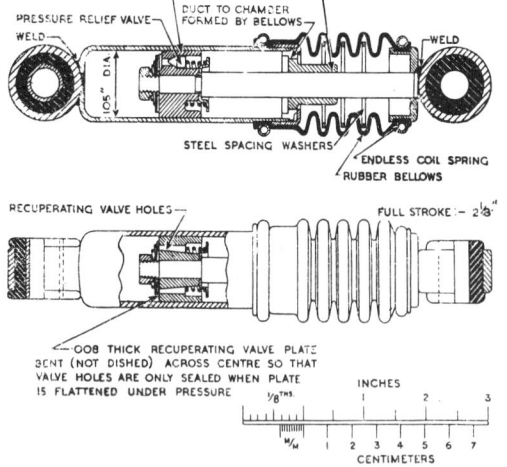

Fig. 30.—Front shock absorber, sectional view

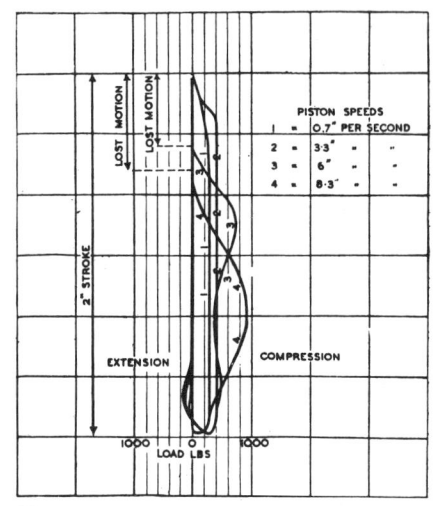

Fig. 29.—Front shock absorber, work diagram

consists of a steel disc held very lightly on its seat by a coil spring.

The piston rod guide is made from cast iron and is held in position by a steel cap which is a press fit on the top of the cylinder. The fixing eyes are welded on, one to the bottom of the cylinder and the other to the piston rod.

The combination of cylinder and rubber bellows is covered by British Patent No. 495621 held by Fichtel & Sachs Aktiengesellschaft.

The fluid capacity of the shock absorber is very small and it was not possible to use any of the fluid for testing, but it is apparently a light mineral oil, very similar to Luvax PF94. Topping up can only be done after removing the bellows from the piston rod.

Operation

The shock absorber is of the single acting type, but is unusual in that the main resistance stroke is on the compression or shortening of the distance between the eyes. Compression of the road spring will be unresisted.

The unit is filled to the top of the cylinder while it is held in the fully extended position. On shortening the shock absorber the piston rod enters the cylinder and oil is displaced unhindered through two large ducts in the piston rod guide into the rubber bellows. Attention is drawn to the recuperating valve plate shown in Fig. 30 which is bent to form a bleed passage for the oil at very slow piston speeds. As the pressure in the working chamber rises, the plate is flattened and the size of the bleed reduced until finally all the oil must pass through the pressure relief valve.

Results

The Work Diagram Fig. 29 shows the resistance obtained at various piston speeds and brings out the fact that at the higher speed, oil does not enter the main resistance chamber quickly enough, and that lost motion takes place at a piston speed of 6 inches per second. This lost motion is about 30% of the total stroke.

REAR SHOCK ABSORBERS—BOGE-ELASTIC

The shock absorbers are of the piston operated, double-acting hydraulic type. The body of the unit is bored to take two pistons, one on either side of a rocker arm which is held centrally between them. The rocker arm pivots about a shaft provided externally with a lever arm. Vertical movement of the car axle relative to the chassis, causes the lever arm to swing through an arc thus rotating the rocker shaft and arm; the latter pushes one piston towards one end of the cylinder and withdraws the other piston from the opposite end of the cylinder. Since both ends of the cylinder are closed, a high pressure is built up in the oil at one end of the unit whilst a decrease in pressure occurs at the other end. A non-return recuperation valve situated in the piston head allows oil to pass from a reservoir to the low pressure end, thus keeping this normal. This leaves the oil pressure on the one piston head opposing lever arm movement. In order to spread the absorption of the shock, a leakage path is provided from each end of the cylinder to the reservoir, thus reducing the rate at which the pressure is built up. Finally, blow-off valves are fitted which limit the pressure reached, thereby controlling the resistance offered by the shock absorber.

When the lever arm is moved in the opposite direction, the function of the two pistons are interchanged. As is frequently the custom with German shock absorbers, the units are arranged to give a low resistance in the direction of the initial shock and a high resistance in the reverse direction to dissipate the energy stored in the springs.

Constructional Details

The components of the shock absorber are shown in Plate 33.

Both pistons are incorporated in a single casting, ground cylindrically at the ends. Some leakage is provided by grooves cut longitudinally in the ground portion. The centre of the casting is cut away to take the rocker arm, which is held between steel pads fitted into holes in the inner faces of the pistons. One pad is spring loaded to take up any possible clearance between the pad and the rocker arm, thus preventing knocking when the stroke is reversed.

The recuperation valves are screwed into the outer faces of the pistons and are off-centre. They consist of a spring-loaded disc seating on an orifice which faces a hole drilled through the piston so as to communicate with the cut-away portion or reservoir. When the pressure in the cylinder drops, the excess pressure in the reservoir pushes the disc from the orifice, thus allowing oil to pass through the orifice, round the disc and into the cylinder.

When the pressure rises in the cylinder it merely seals the orifice more effectively.

A single drop forging forms the rocker arm and shaft.

The ground shaft is supported at one end in the body casting and at the other end by

Plate 33.—Rear shock absorber, exploded view

a cast iron bush with brass lever. A rubber gland seals the bearing at the lever arm end whilst a steel plate and fibre gasket cover the other end. The lever arm—7⅝ ins. long—is held on the shaft by a self-cutting spline.

The shock absorber body is an iron casting, 7 ins. × 4½ ins., overall size. The cylinders are bored from one end, the opposite end being closed. The open end is sealed by a dish-shaped disc and rubber washer, which are held in position by a hexagon-headed cover. A tapped hole is provided in the top of the body for filling and topping-up, and this is normally sealed by a round-headed screw and fibre washer.

The arrangement of the blow-off valves is shown below. "A" and "B" are holes drilled through from the ends of the body to the valve housing. These holes are drilled before the cylinders are bored, and in the

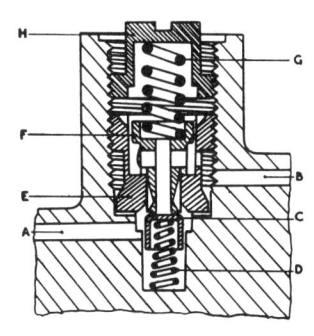

Fig. 31.—Rear shock absorber valve, section view

case of the closed end a brass plug is inserted to form a seal. "C" is a cup loaded by a helical spring "D." The slides of the cup are slotted to allow free circulation of the oil in the lower chamber. The cup presses against the bottom of the valve "F" and in its normal position seals the bolt through the centre of the valve "E." The valve housing is screwed into the body by means of slots milled in the top and sealed by a fibre washer. The valve rests on a shoulder of the housing so that in its normal position oil can pass through from the lower chamber only by means of a leakage path provided by two slots milled in the valve seating. Two holes in the valve walls allow oil to pass from the centre to the outside of the valve and three holes in the valve housing allow oil to circulate between this point and hole "B." The valve is loaded by a spring "G," which locates at opposite end in a screwed cap "H," the position of which can be adjusted by means of a slot in the top. The whole is finally enclosed by means of a hexagon-headed cover.

When the pressure commences to rise in "A," oil passes through the leakage paths into the upper chamber and thence through the holes in "E" and out through "B" to the other side of the shock absorber. When the pressure in "A" rises above the desired maximum, the pressure on the underside of "C" is sufficient to overcome the spring "G" thus causing the valve "F" to lift. This allows the oil to pass rapidly round the valve, relieving the pressure and preventing any further increase above the desired maximum. When the pressure in "B" is rising, oil first leaks into the same path as for "A," but in the reverse direction. Oil also circulates into the centre of the valve and when the desired maximum pressure is exceeded it presses the cup "C" away from the valve against the pressure of the spring "D," thus allowing rapid leakage past this point and relieving the pressure.

The valve gear is adjustable. By varying the slots cut in the valve seating the leakage can be adjusted so as to vary the rate at which pressure is increased. The compression of the spring "D" can be adjusted by turning the cap "H," thus varying the blow-off pressure for the cylinder to which this is connected. Variation of the blow-off pressure for the other cylinder can be effected only by the choice of the spring "D."

Test Results

The shock absorber was tested on a work diagram machine which gives torque in inch-pounds against a function of the vertical movement of the lever arm. Speeds of 20, 100, 180 and 250 degrees per second were used and the resultant diagrams are attached. From these it will be seen that the resistance is exerted principally with downward movement. At the standard test speed of 100° per sec. a maximum torque of 1,700 in.-lb. is obtained, while 250 in.-lb. is the maximum reverse torque neglecting the inertia oscillations produced by the testing machine.

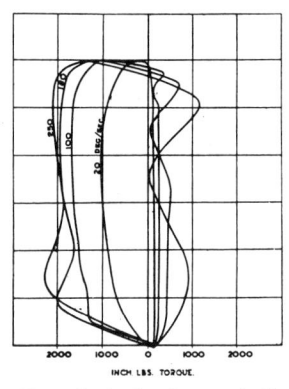

Fig. 33.—Rear shock absorber, work diagram

Viscosity and pour point tests were made on a sample of the oil; these results being as shown on the accompanying curve, together with a curve of the oil used in "Luvax" piston type shock absorbers. It will be seen that the German oil has a higher viscosity than the "Luvax," whilst in addition the pour point is only —15° F. compared with —47° F. for the "Luvax" oil. This would prevent the unit from being used in cold climates.

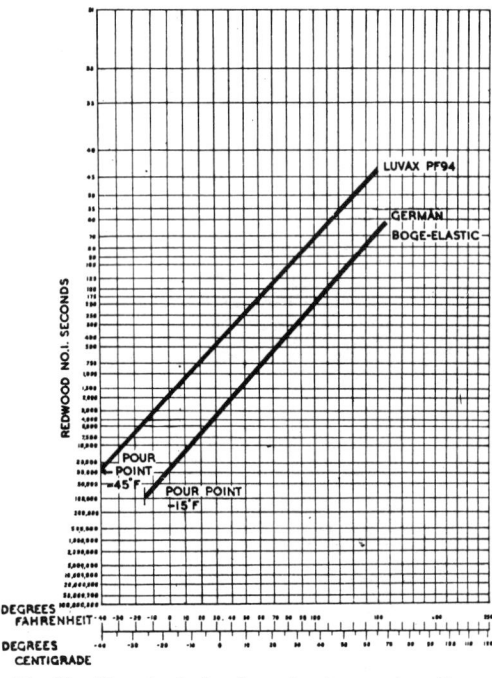

Fig. 32.—Rear shock absorber valve, temperature diagram

STEERING

General Description

Steering control conforms with Continental practice, being on the left-hand side of the vehicle. The steering gear consists of a separate steering box which is connected to the inner column through a fabric universal coupling joint at the lower end, thus insulating the handwheel from road shocks. The steering box is rigidly clamped to the upper torsion-bar cross tube on the front end of the chassis under-frame. The track rod is divided, giving independent steering as well as suspension, and adjustment is provided for setting the track.

Track Rod System

The ball joints of the track rod system are positioned so that correct geometry is maintained during the bump and rebound movement of the wheel. For example, the wheel travels up and down through an arc of radius equal to the suspension link, causing the outer steering arm ball joints (which are attached), to follow the same path. The inner track rod ball joints, attached to the steering box lever, are positioned relative to the outer joints so that the track rods generate a cone whose apex is at the inner joint. Thus directional control of steering is accurately maintained.

Turning Circle

The turning circle, ascertained when the vehicle was first received, is 30 ft. 5 ins. on right-hand lock and 36 ft. 8 ins. on left-hand lock. The left-hand lock was limited by the steering box stop and the right-hand lock by the tyre fouling against the suspension link. The uneven locks were due to the steering column location in the body, necessitating the steering unit being lined up in an incorrect position on the chassis cross tube. This in turn causes the rocker lever in the steering box to be out of the central position for straight ahead driving. Since neither the body nor the chassis appear to have been damaged to any great extent, it would seem that this is bad design, especially in view of the absence of locating means for the steering box on the cross tube. It was ascertained that when the steering box is correctly positioned, both locks are limited by the tyres fouling the suspension links. With the narrower tyres originally specified for the Volkswagen, the locks were restricted by the stops in the steering box, which also is bad practice in view of the fact that thrust loads on full lock are transmitted through the steering box.

Description of Steering Box

This is of the worm and rocker arm type (with segment nut interposed), and has a low efficiency. Its steering ratio characteristics provide a lower ratio in the straight ahead position than on lock. The design is unusual, being as follows: A stem having a thread similar in form to that of the Acme, operates a segmental nut provided with four threads and having approximately one-third circumferential engagement. The nut is hemispherical in shape and fits into a correspondingly shaped seat on the rocker shaft. The nut acts as the medium through which rotation of the stem causes angular displacement of the rocker shaft, and the nut oscillates in the seating during alternating locks. The nut therefore moves around the thread, contact between the two being maintained by a spring and causing the rocker shaft to move endwise. It was observed that the nut had been over-riding its seating, causing considerable oval wear in the latter. This may be attributed to neglect in adjusting the spring when end play occurred, or alternatively the spring pressure may have been insufficient to withstand the thrust loads experienced on this vehicle.

This design of steering gear follows the principle described in Patent No. 384195 applied for by Wanderer Motor Co. and F. Porsche in 1931. It is claimed in this that the spring which presses the nut against the thread eliminates the shock loading from the hand wheel.

CONSTRUCTIONAL DETAILS

Steering Box Unit

This is a malleable iron casting having an integrally formed half-round mounting, the corresponding clip being a pressing. The rocker shaft is supported directly in the casing and a rubber oil seal is fitted at the lower end. The stem is supported on self-contained cup and cone type ball races, the

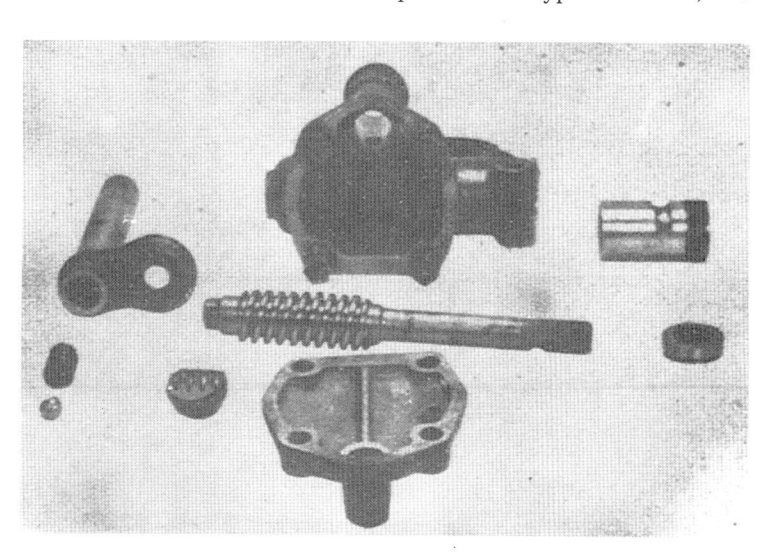

Plate 34.—Steering unit, exploded view

method of adjustment for wear being noteworthy, as follows:—A sleeve having a slow helix angle groove cut on the outside diameter fits into the casing and is held by a pinch bolt. When the latter is loosened and the sleeve is rotated, it is compelled to move endwise by the pinch bolt, thus causing the bearing cup to move and take up end play. The stem is made of hardened steel, has a fixing groove milled at one end and fine serrations rolled at the same end. This provided the means of attachment for a fabric coupling which is connected to the steering column in a similar manner.

A magnesium base alloy die cast steering box cover is used; this incorporates an adjuster for the rocker shaft spring and a filler plug.

Steering Column

A tubular inner column is fitted, split at the lower end for clamping on to the fabric coupling flange and having an adaptor welded on the top end, with coarse serrations for the handwheel fixing.

Steering Wheel

The steering handwheel is of the three-spoke, solid type, having a composite construction with plastic base black covering. The hub is of aluminium base alloy and made by Petrilenkrad of Aschaffenburg.

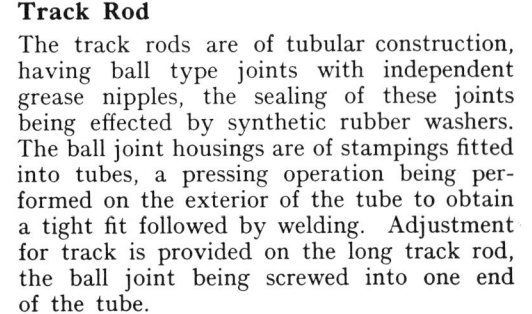

Track Rod

The track rods are of tubular construction, having ball type joints with independent grease nipples, the sealing of these joints being effected by synthetic rubber washers. The ball joint housings are of stampings fitted into tubes, a pressing operation being performed on the exterior of the tube to obtain a tight fit followed by welding. Adjustment for track is provided on the long track rod, the ball joint being screwed into one end of the tube.

The design of the ball joints could not be ascertained without risk of damaging the seal provided by a Welch type plug. Since the track rods are formed in one piece with the joints by the method previously described, this entails replacing the complete rod when the joints have worn excessively. It was

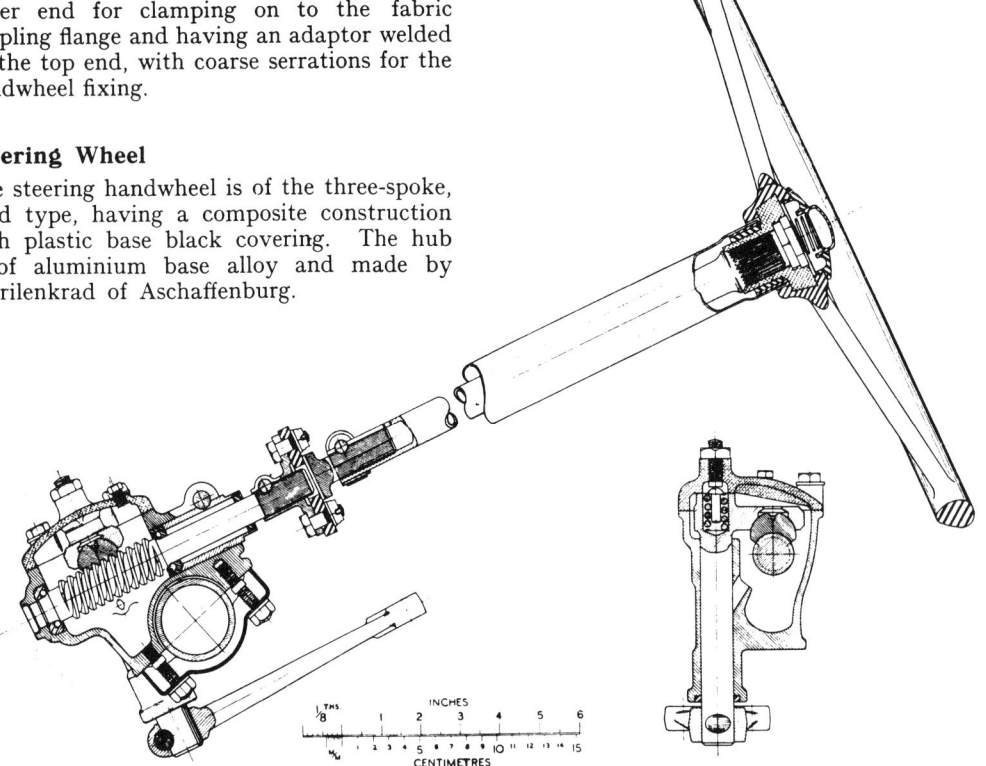

Fig. 34.—Steering box and column, sectional view

The inner column is enclosed in a thin section tube which is supported at each end in rubber grommets on the body dash and panel rail. The handwheel hub is supported on a plastic base bush fitted in the tube.

apparent that the joints have spherical seatings and the ball pins have taper end fitting, the joints being upcast on all the levers. Lubrication is arranged by the provision of grease nipples.

STEERING DATA

Overall Ratio (straight ahead): 15·7 to 1.
Overall Ratio (at R.H. full lock position): 13·9 to 1.
Overall Ratio (at L.H. full lock position): 11·7 to 1.
Handwheel movement (overall): 2·75 turns.
Handwheel movement (central to R.H. full lock): 1·5 turns.
Handwheel movement (central to L.H. full lock): 1·25 turns.

Turning Circle (R.H. lock): 30 ft. 5 ins.
Turning Circle (L.H. lock): 36 ft. 8 ins.

Rocker Arm

Centres: 1·4 ins.
Shank Dia.: 22 mm. (·87 in.).
Shank Length to Steering Lever: 4·75 ins.

Stem Thread

Top Dia.: 27 mm. (1·06 ins.).
Bottom Dia.: 20 mm. (·79 in.).
Number of Starts: 2.
Lead: 12·5 mm. (·49 in.).
Pitch: 6·25 mm. (·245 in.).
Angle between sides of thread 18° included.

Handwheel

Outside Dia.: 15·50 ins.
Serrated Fitting: 24 mm. (·95 in.) O/dia. shaft; 22 mm. (·87 in.) I/dia. hub; 24 serrations parallel.

Inner Column

22 mm. (1·06 ins.) O/dia. × 16 mm. (·63 in.) I/dia. × 37 in. approx. length.

Cup Ball Race

32 mm. (1·26 ins.) O/dia. × 10 mm. (·39 in.) wide.

Type F — 47 — 05.

Outer Tube

40 mm. (1·57 ins.) O/dia. × 1 mm. (·04 in.) thick.

Track Rod (long): 18 mm. (·71 in.) O/dia. ball joint centre 31 ins.

Track Rod (short): 16 mm. (·63 in.) O/dia. ball joint centre 13·75 ins.

BRAKING SYSTEM

The braking system provides for both foot brake and hand brake operation on all four wheels, the same cable gear being utilised in each case. Provision is made for the hand brake and foot brake to operate independently of one another despite the use of the same cable gear. The cable operates through guide tubes and flexible conduit on the non-compensated system. Two-shoe internal expanding brakes are used, cheapness of design and construction appearing to be the principal objects.

BRAKE SHOES
General Description
Two-shoe type internal expanding brake shoes are used, operated at the tips by a floating type of mechanism. The latter consists of a bell crank lever so arranged that the operating cables are at right angles to the brake back plate. Adjustment for shoe wear is provided opposite the shoe tips, i.e., at the fulcrum or anchorage. Malleable iron castings are used for the brake drums.

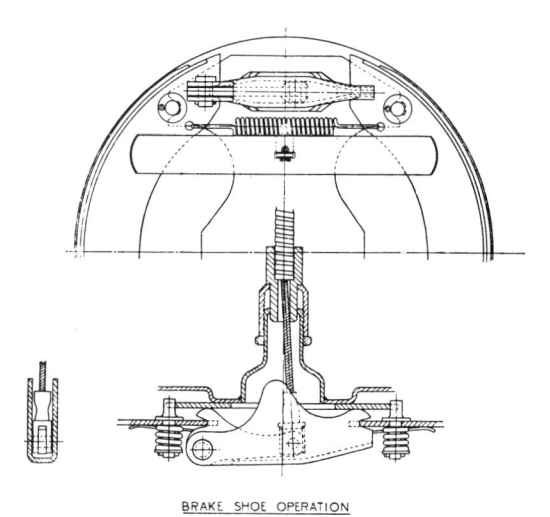

Fig. 35.—Brake shoe operation

Plate 35.—Brake shoe assembly, L.H.

Floating Operation
This is shown in Fig. 35, and consists of a steel bell crank lever attached to a pressed steel housing, forming a self-contained operating unit. The operating unit is located in notches cut in the shoes and retained by the shoe return springs.
The whole is arranged to float and thus equalise the force applied to the shoes. The effort is applied at the operating lever, involving a high unit pressure at the reaction points between the operating unit and the back plate. Since sliding or wiping takes place during brake operation, a high frictional resistance occurs, thus impairing the efficiency.

Adjuster
The adjuster mechanism is arranged in a steel stamped housing which also serves as an anchorage and is welded to the back plate. Shoe adjustment is performed by rotating a hexagon-headed screw inside a threaded sleeve in the housing; this sleeve is compelled to move axially being held against rotation by a split pin fitting in a slot. Movement for the adjustment of clearance is conveyed to the shoes through hardened steel plungers, which fit in the housing and have chamfered ends for engagement with a shoulder on the sleeve. The adjusting screw is locked in position by means of radial serrations locating in corresponding serrations formed on the brake back plate and held in contact through the pressure exerted by the brake shoe return springs.

Shoe Linings
Moulded linings having an asbestos base with copper content are riveted to the pressed steel shoes.

Brake Drums
The brake drums are integral with the hubs and strengthened by ribs on their outer periphery. Water-excluding means is provided by fitting a deflector cover, which is clipped on to the outside of the back plate.

Other Details
The brake back plates are steel pressings, the fronts not being interchangeable with the rear owing to differences in fixings, etc. Reinforcement plates are welded on, providing rigidity for the fixing, operation point and anchorage.

BRAKE CONTROLS
General Description and Construction
As previously mentioned, the foot brake operates the brakes on all four wheels. The controls are mainly cable and the system is non-compensated. The brake pedal forms part of a sub-assembled unit comprising all pedal controls—i.e., accelerator and clutch—which is bolted to the underframe backbone and constitutes a rigid anchorage. The pedal

is a steel stamping, keyed and bolted to the operating lever shaft; this consists of a steel stamping having the shaft and lever formed integrally. The pedal assembly fulcrums on self-lubricating bushes, and these are offset in relation to the pad. The front of the backbone is open-ended to receive a control head which forms the point of attachment for the four cables. The cable end ferrules are located in the control head, which is a steel stamping and is welded to the end of a channel member. This forms the compression member which is operated by the pedal and is slotted at the rear end attachment to permit independent operation of the hand brake. The front end of this member is supported in a pressed steel strip on which it slides during operation. The control head is also connected to a compression rod from the hand brake, and this fits snugly inside the channel member locating on a spherically ended lever, bearing in a hole in the end of the rod. The latter is also slotted to permit independent operation of the footbrake.

The compression members are returned to their original positions after operation by means of a coil spring situated at the front end of the backbone.

The cables are completely enclosed in rigid tubing and flexible conduit; the rigid tubes being welded to the backbone act as guides. No arrangement for greasing is provided, and there is also no provision for adjusting the cable for stretch, except for a small amount on the back plate abutment for the conduit. This latter adjustment appears to be sufficient only to cope with possible variations in production. Owing to the disposition of the control head there is great disparity between the lengths of the front and rear cables.

HANDBRAKE UNIT
General Description and Construction

The handbrake is arranged in a horizontal position and is fitted between plates welded on top of the backbone, which provides a rigid mounting. The handbrake lever is a folded steel pressing and fulcrums direct on a hardened steel pin, normal type ratchet and pawl (with concealed plunger rod operation) being fitted. The whole unit forms an independent assembly and is cheaply constructed. Further observation revealed that the pawl was produced from the ratchet pressing, thus obtaining both cheap and light construction.

CONTROLS
Brake, Clutch, and Accelerator Pedal Assembly Unit

This assembly consists of a housing made from a steel stamping and tubing welded together. The former is arranged to carry the accelerator pedal from an independent fulcrum and also forms the mounting face to the chassis. Both the brake and clutch pedals are attached to shafts carried in self-lubricating bushes, and it was noted that the clutch shaft fitted inside the braked shaft. This is not generally considered good practice owing to the possibility of one shaft sticking or seizing inside the other, in point of fact the brake pedal did not operate very

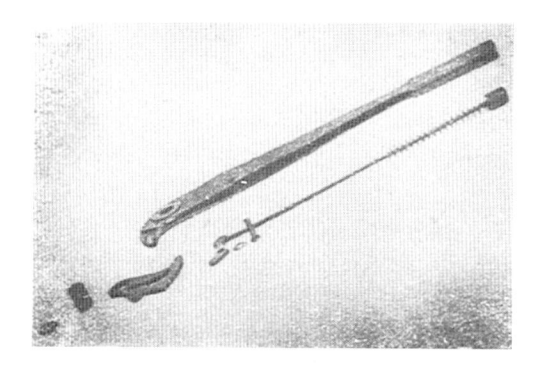

Plate 36.—Handbrake, exploded view

freely on the shaft owing to the accumulation of dirt. The accelerator pedal is a flat steel pressing and has a bobbin or roller type operation, produced from plastic base material. The accelerator pedal is connected to the carburettor by piano wire operating inside steel tubing which is located inside the backbone.

The choke control is arranged similarly, but is hand operated by a knob situated on top of the backbone.

The clutch control is operated by means of stranded wire, directly connecting the pedal to the operating lever at the rear. The wire is adjustable and fits inside a tube housed in the backbone.

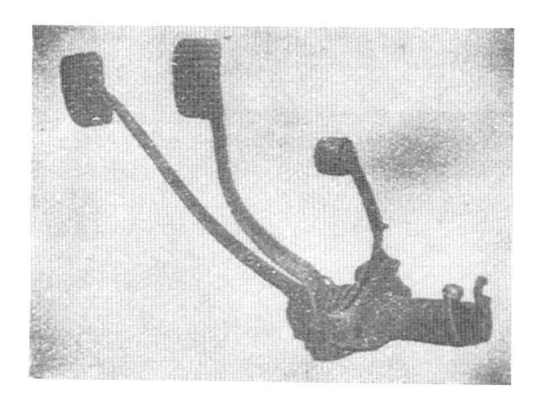

Plate 37.—Pedal unit

46

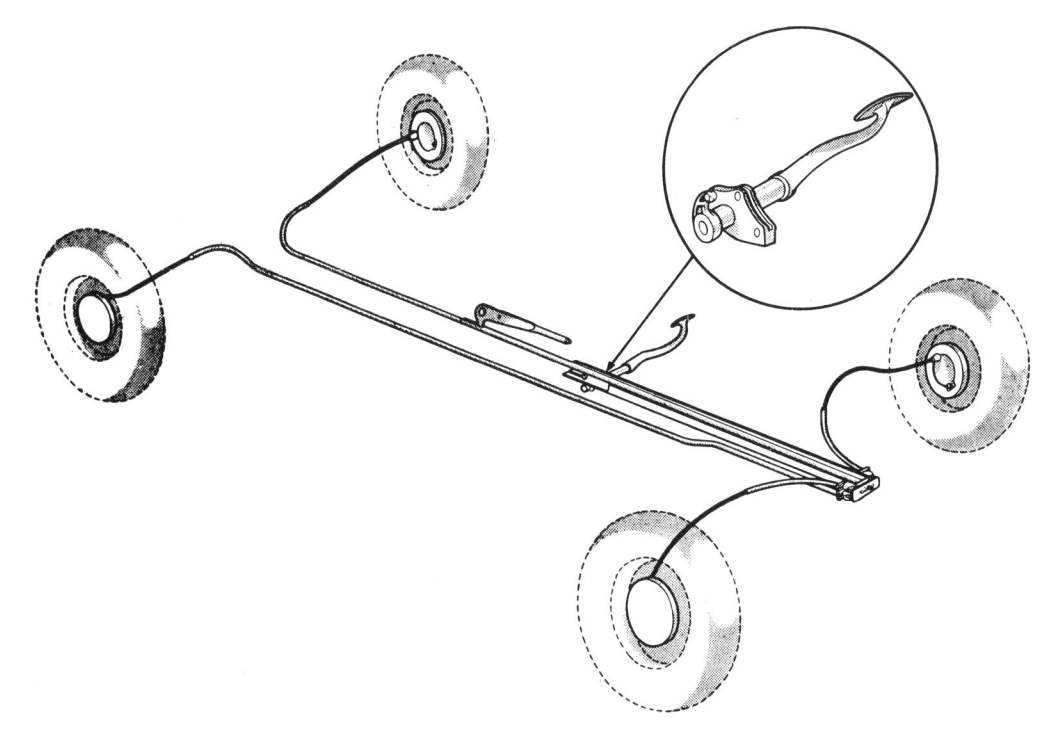

Fig. 36.—Footbrake and handbrake controls

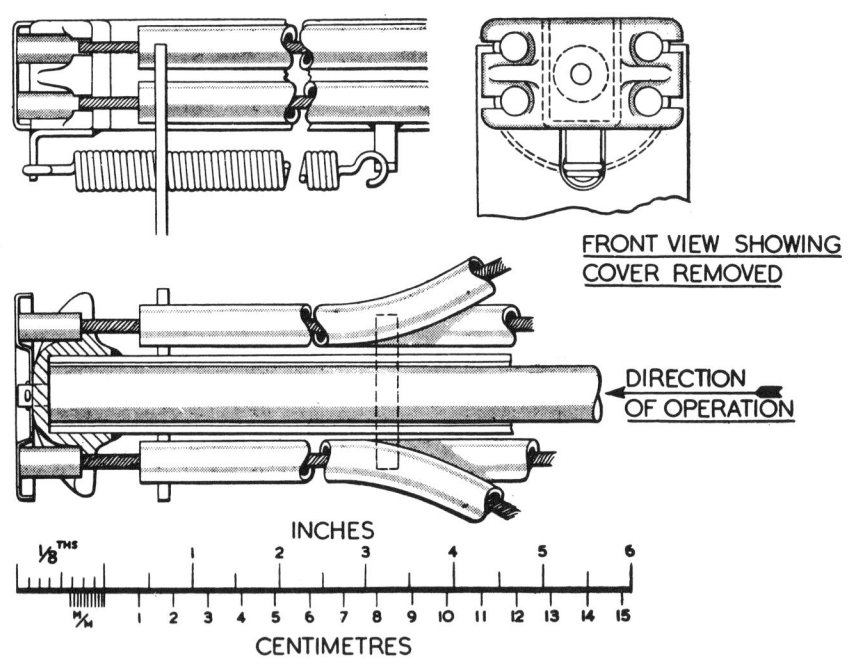

Fig. 37.—Brake control head

BRAKES—General Data

Brake shoe floating bell crank leverage:
$\dfrac{50 \text{ mm.}}{14 \text{ mm.}} = 3 \cdot 57 : 1.$

Pedal leverage (to bell crank lever):
$\dfrac{215 \text{ mm.}}{25 \text{ mm.}} = 8 \cdot 6 : 1.$

Pedal leverage (to shoe tip): $= 30 \cdot 7 : 1$ overall.

Brake power ratio: 50% front and 50% rear.

Handbrake leverage (to bell crank lever):
$\dfrac{305 \text{ mm.}}{41 \text{ mm.}} = 7 \cdot 43.$

Handbrake leverage (to shoe tip) $= 26 \cdot 5 : 1$ overall.

Brake drum (inside dia.): 9 in.

Brake shoe lining (width): 1·15 ins.

Brake shoe lining (effective angular contact): 116° per shoe.

Brake shoe lining (effective area): 84·2 sq. ins. total

Laden load per sq. in. lining: 30·75 lbs.

Diameter of brake cable (stranded wire): 0·14 in.

Handbrake unit weight: 1·125 lb.

Brake, clutch and accelerator pedal unit: 5·47 lb.

WHEELS AND TYRES

WHEELS
Size: 4·25 × 12.

Type
Flat base rim type of two-piece pressed construction. The nave is integral with one half of the rim. The other half of the rim is in the form of a flange bolted to the upper periphery of the nave. (See drawing.)

Hub Fitting
Five-stud of conventional Continental type.

Weight
12¾ lb.

TYRES
General Description
The tyre used is apparently a plain tread aeroplane tyre as listed in German data books. Its adoption on a wheeled vehicle has presumably been governed by considerations of tractive effort on loose sandy surfaces and under such conditions it should give quite satisfactory performance. Tread life, under even intermittent road work, would be very small. None of the tyres appears to have worn much and no signs of uneven wear were evident. The spare wheel had not been used.

Construction
The tyre is of the ordinary pneumatic type and embodies no bullet proofing features. The tube is conventional, and is fitted with a rubber covered valve singly bent to suit the wheel. Synthetic rubber is used for the tread and side walls of the cover.

Nominal Size Marking
690 × 200 (8·00 — 12).

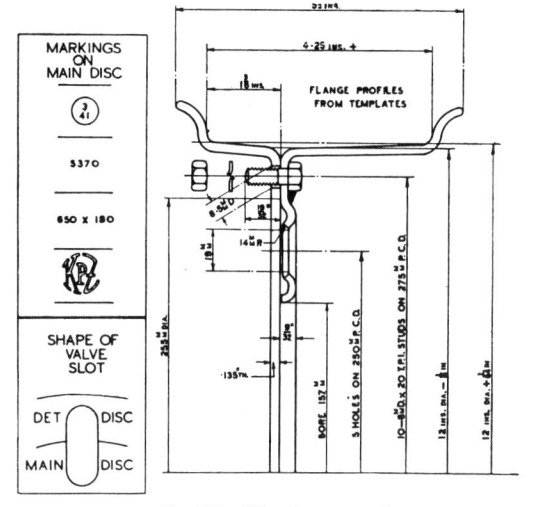

Fig. 38.—Wheel, cross section

Manufacture
Continental.

Side Wall Markings
45415482 B.S. PF1. A1.

Inflated Dimensions
Overall Diameter: 26·4 ins.
Overall Width: 7·6 ins.
Tyre Pressure estimated: front, 15 lbs. per sq. in.; rear, 18 lbs. per sq. in.
Effective Rad. measured: 12·3 (static laden condition)

Tread Pattern
Plain.

Weights
Cover, 20½ lb.; tube, 3¼ lb.

Rim Fitment
Rim Section: 4·25 ins. Rim dia.: 12 ins.

LUBRICATION

The lubrication system is arranged for manual operation through the usual type of grease nipples, the design being such that these are reduced to a minimum. In some cases grease points are also arranged where bushes are made from plastic base material. The tool kit (including the grease gun) was missing and, therefore, the type of gun used is not known. The oil capacities of the various units and also the types of oil used for refilling are as follows:—

Engine Sump: 3¾ pints
Engine Air Cleaner: ¼ pint.
Gearbox and Rear Axle: 1 gall. Triple Shell (Heavy).
Rear Axle Reduction Boxes: ¼ pint each. Shell Spirax Gear Oil.
Steering Box: ⅓ pint. Shell Spirax Gear Oil.

BODY-CONSTRUCTIONAL DETAILS

An open type tourer body is fitted, which is almost entirely fabricated from flat steel panels of approximately 20 S.W.G. (·036) thickness, ribbed for stiffening.

A collapsible cape cart type of hood is employed and is covered with woven fabric. Removable side curtains with celluloid windows of conventional type are also provided. The hood frame is of tubular steel and the fabric cover is secured to the body by means of straps passed through loops welded to the body.

Seating is arranged for three persons; two front seats and one rear seat on the left-hand side. Tubular steel is used for the seat frames, the front being of the "Bucket" type with coiled steel tension springs for cushioning effect. The seats are covered by thin wadding pads stitched between hessian canvas inside and enamelled duck on the outside. The tubular top rail is also used as a brace across the body between the B and C posts.

The space normally occupied by the right-hand rear seat is used to accommodate a steel trunk, access to which is gained by opening the right-hand rear door, which is provided with a rubber seal to prevent ingress of water and sand. Behind this and the rear left-hand seat is a second trunk which is opened at the top, using a lid of precisely the same size and shape as the rear engine compartment lid, but having a small extension piece at the front to form a water drip. On top of this lid are felt-lined crates, presumably for carrying batteries for wireless purposes. It would be impossible to open this lid when the batteries are in position.

The windshield is designed to fold forward to an approximately horizontal position; spring clips (with rubber cushions) are arranged on the scuttle at the front to hold the windshield in the down position.

The spare wheel is mounted on the front scuttle, which is suitably reinforced at this point, and has a semi-tubular column running from the dash to the extreme front end. This column can accommodate tools or miscellaneous articles, the dash end being left open.

The doors and locks are designed so that all four doors (which are hinged on the centre pillar) are interchangeable, but the right-hand rear door is provided with extra members to accommodate sealing rubbers for the trunk already mentioned.

The engine compartment at the rear is divided from the body proper by a sheet metal wall, stiffened with ribs and V-section reinforcements. The lid is fitted with a piano-hinge and a stay to hold it in the open position. The walls of the engine compartment carry brackets of various designs to carry tools, spare oil, cans, etc., the bottom being open to accommodate the engine. The joint between the engine compartment and the floor is effected by the forked rubber sealing strip shown in the illustrations (Figs. 50 & 51).

Slots are provided in the rear of body for permitting entry of air into rear compartment for cooling engine.

No padding or trim pads are employed except on the seat and battery crates as already mentioned. The same paint finish is used for both inside and outside.

UNDERFRAME-CONSTRUCTIONAL DETAILS

The underframe is designed to utilise sheet metal wherever possible.

The main floor is of sheet metal, approximately ·048 in. thick and ribbed for stiffening, whilst down the centre runs a semi-tubular "backbone" approximately ·098 in. thick. This backbone is arranged at the front to carry the torsion bar tubes, and at the rear is forked to form the engine cradle.

Mounted on the backbone is the handbrake lever, the latter lying in the horizontal plane when in the "off" position. The bottom of the backbone is closed by ribbed plate spot-welded in position.

Foot pedals are carried on the side of the backbone and are extended by means of cranked levers to a position convenient to the driver.

The body structure is attached to the underframe by means of bolts, a continuous rubber insulating strip being employed between the two.

The main outside joints are of the clinched flange type. Elsewhere spot-welding is employed throughout, reinforced with occasional gas-welding and rivets, where access is difficult for welding dies, or where additional strength is required.

The petrol tank is suspended in the right-hand side of the dash and is fabricated from sheet steel, roller-welded at the joints. Leaks in the joints have been filled with solder.

The vehicle has been repaired at some points, i.e., three door hinges and the handbrake lever have been broken, repairs being effected by gas-welding.

In addition a hole has been cut in the backbone to service the foot pedals; this hole has been left open.

The main floor is covered with wood-slat staging.

Approximately 100 lb. of sand was found inside the vehicle, some of which had accumulated in the hollow section box members.

Covering the front torsion bars, steering gear, etc., is a sheet steel shield; this is omitted from the illustrations for clarity.

The rear seat cushion was missing from the sample vehicle.

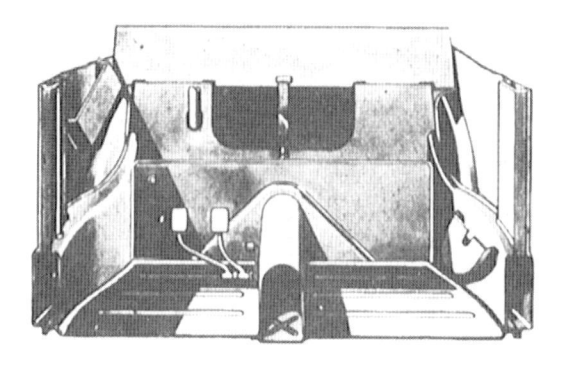

ARRANGEMENT OF DASH STRUCTURE
(PETROL TANK OMITTED)

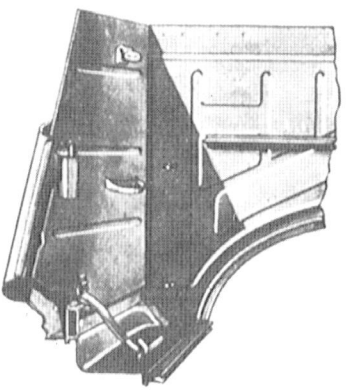

CORNER OF REAR ENGINE COMPARTMENT

PETROL TANK (MOUNTED IN DASH)

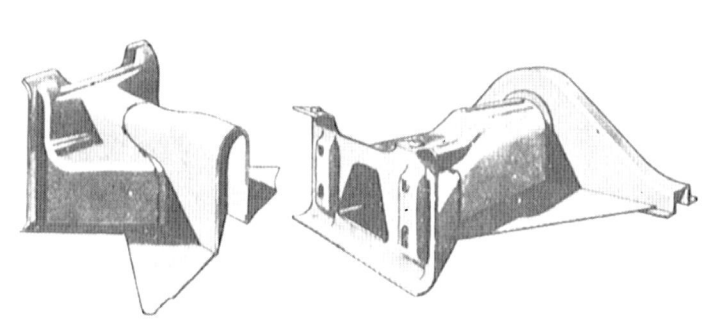

FRONT END OF BACKBONE

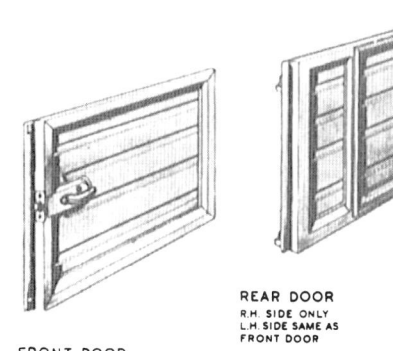

FRONT DOOR

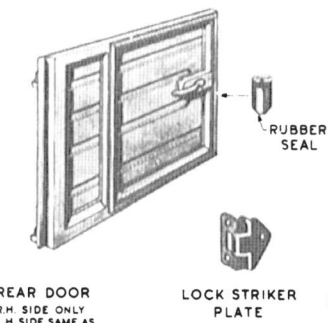

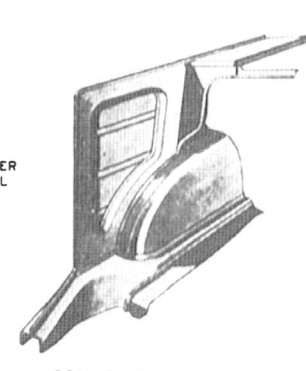

REAR DOOR
R.H. SIDE ONLY
L.H. SIDE SAME AS FRONT DOOR

RUBBER SEAL

LOCK STRIKER PLATE

BODY-REAR QUARTER

RIFLE CLIP ON R.H. FRONT WING

WINDSHIELD (DOWN POSITION) CLIP ON FRONT SCUTTLE

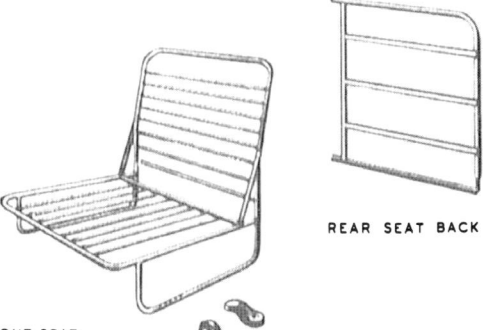

FRONT SEAT

SECURING CLAMP FOR FRONT SEAT

REAR SEAT BACK

HAND-BRAKE LEVER

SHOVEL SLING ON R.H. SIDE OF DASH

Fig. 39.—Body details

50

Fig. 40.—Body details, sectional views

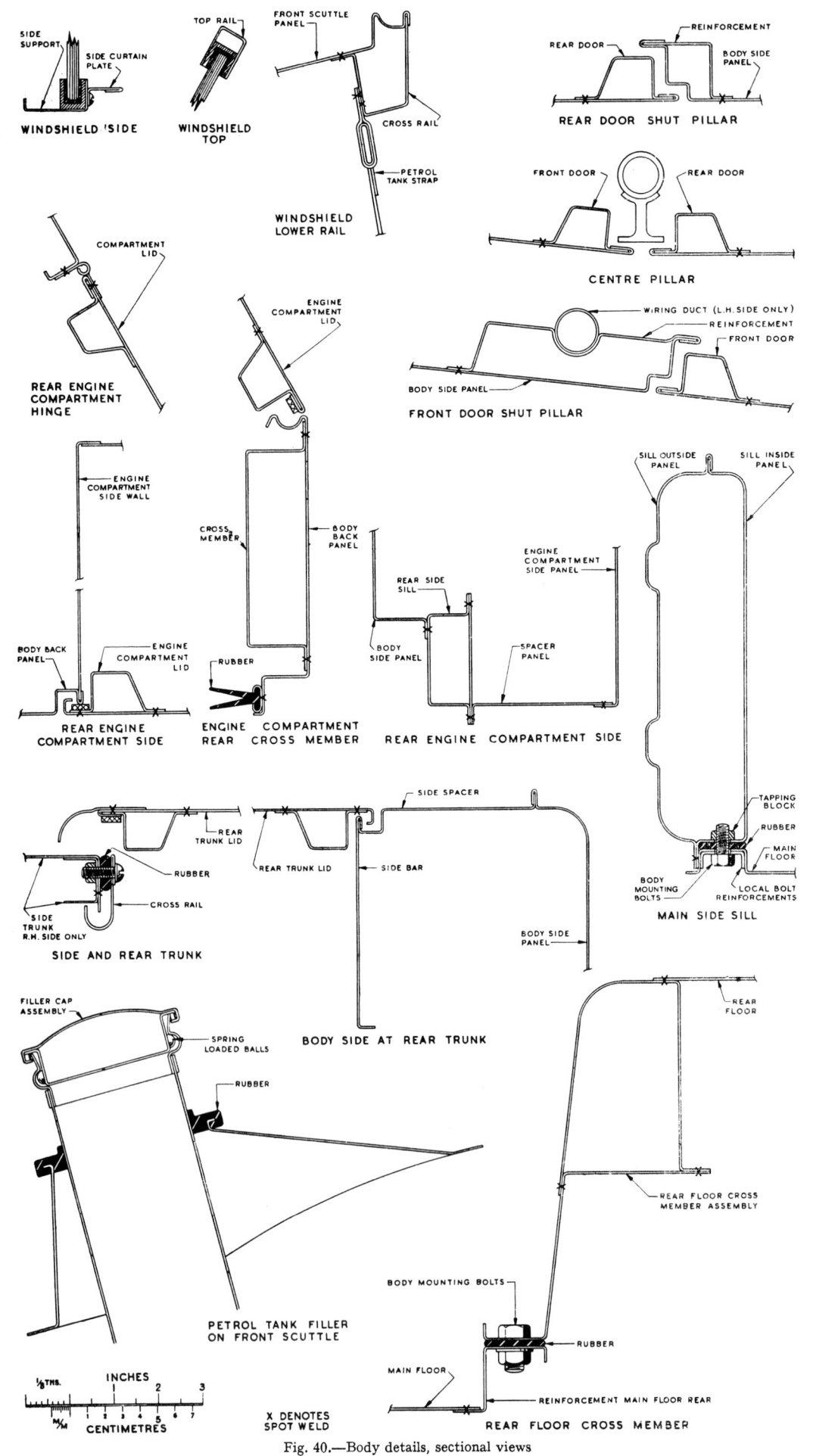

GENERAL DATA

Weight—Body : 504 lbs.

This included wings, headlamps (but not side lamps), collapsible hood, windscreen trafficators, floor (rear of "heelboard"), steering wheel and box.

Weight—Underframe and Floor : 196 lbs.

This includes main floor and foot pedals, change-speed lever, handbrake lever (with control links running through backbone) and tube for rear torsion bars.

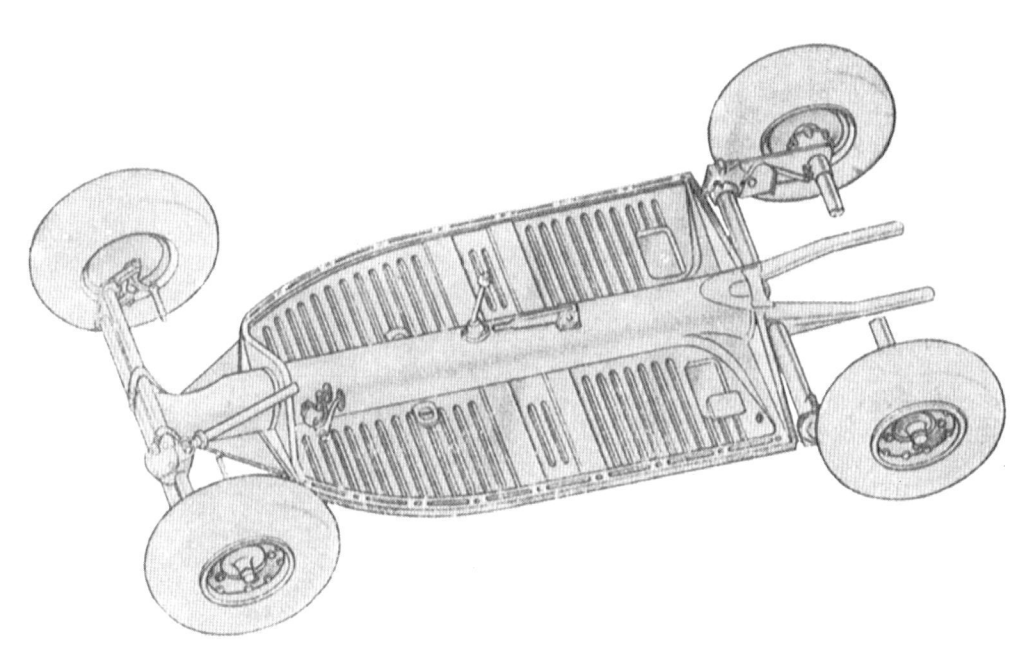

Fig. 41.—Underframe, top

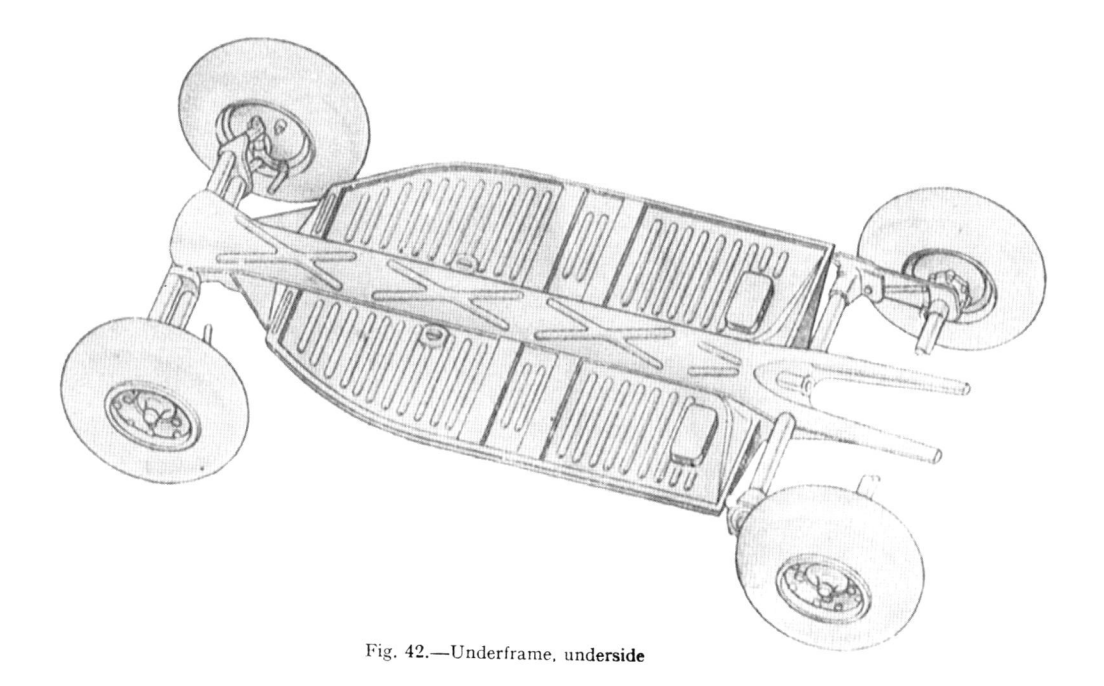

Fig. 42.—Underframe, underside

ELECTRICAL EQUIPMENT

IGNITION EQUIPMENT

The ignition equipment comprises a Bosch distributor, Type VE4BS276, Bosch ignition coil, Type TL6, and normal rubber insulated ignition cable fitted with suppressors at the plug ends.

Distributor, Type VE4BS276

The distributor is of the four-cylinder type, fitted with a single contact breaker and provided with an automatic timing control of the usual centrifugal type.

The distributor body is of cast iron $2\frac{9}{16}$ ins. diameter. The shaft runs directly in the cast iron without bearing bushes. No means of lubricating the bearing is provided—this is presumably dependent upon splash from the engine. An oil-excluding thread is cut at the top of the shaft to prevent the ingress of oil into the distributor body and the shaft bearing is relieved for approximately $\frac{5}{8}$ in. at its centre. Experience has shown that cast iron distributor bearings are prone to seize unless positively lubricated, and it is likely that trouble may occur in service due to this.

The automatic advance mechanism is of the usual Bosch design as shown in Plate 38

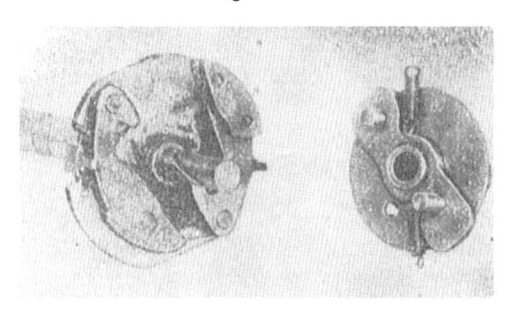

Plate 38.—Distributor automatic advance mechanism

The short projections on the eights at the pivot end act directly on pins attached to the cam foot, control being applied by two springs which are extended round the curved faces of the cam foot as the cam advances, and this produces a kinked curve characteristic. Both control weights are in the form of single laminations and the cam foot is attached to the cam body by being forced on, the cam body afterwards being lightly riveted over. A circular plate is clamped between the cam foot and the cam body and projections raised on this plate form anchorages for the springs, the other ends of which are attached to the upturned ends of a steel strip screwed to the underside of the action plate. Slots are provided in the strip for adjustment

Plate 39.—Distributor, showing contact breaker

purposes and the spring pillars form the stop for the weights. The range of advance is limited by one of the cam pins which is longer than the other and operates in a hole in the action plate. The cam bearing is $1\frac{1}{32}$ ins. in length and the shaft is relieved in two places by grooves $\frac{3}{32}$ in. wide. The cam and shaft are unhardened; the former showed distinct signs of scoring. The control weights and bearing pins are hardened and the latter are secured to their respective components by having metal forced into grooves in the pins. A curve of the automatic advance characteristic is attached.

The distributor shaft is driven by an offset dog which forms a semi-universal joint. The dog is secured to the shaft by a pin $\frac{5}{32}$ in. diameter, the pin being a drive fit in the shaft and a slack fit in the dog, which is bored out larger than the shaft diameter. This allows the dog to be self-aligning in a direction at right angles to the dog. The pin is retained in position by a circular spring wire fitted into a groove in the dog and over the ends of the pin.

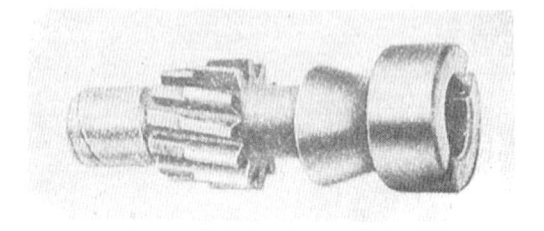

Plate 40.—Distributor driving shaft

The metal contact breaker plate is pierced with six holes of $\frac{7}{16}$ in. diameter in order to reduce the weight. Contact breaker adjustment is made by a screw having an eccentric head which moves the stationary contact plate between predetermined limits. The contact breaker lever is of the lightened type, having a blued steel spring riveted to it; the same rivet is also used to secure the bakelised fabric heel. No cam lubrication is provided. The tungsten contacts are brazed directly on to the steel lever shell and on to the angle plate.

The condenser is of conventional construction and has the live connection—a round wire enclosed in insulating tubing—brought through a bakelised fabric washer and connected to the low tension terminal of the distributor. The condenser securing clip is projection-welded to the steel case.

Plate 41.—Distributor, showing H.T. wires

The rotor arm is of circular shape, the brass electrode being moulded in.

The distributor moulding is of the horizontal outlet type with a loose cover and a shroud to protect the leads from water where they emerge from the moulding. Grooves are moulded in the distributor to take the cables, which are pierced by studs fitted in the distributor.

The track diameter is 43 mm. and the material used appears to be wood-filled bakelite, since it tracks very readily on 2 mm. and 8 mm. gaps.

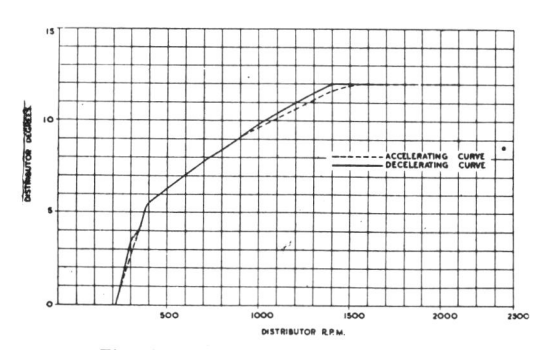

Fig. 43.—Distributor, distribution curve

Further details of the distributor are given below:

Weight of complete distributor	2 lb. 4 oz.
Shank length (under lever)	1 $\frac{11}{32}$ ins.
Shank diameter	1·062 ins.
Shaft diameter	·490 in.
Shaft protrusion	$\frac{21}{64}$ in.
Dog width	·176 in.
Dog diameter	·984 in.
Dog bore	$\frac{17}{32}$ in.
Dog protrusion	·532 in.
Clamping plate thickness	·102 in.
Outside diameter of moulding	2 $\frac{15}{16}$ ins.
Track diameter	43 mm.
Distance between electrodes	25 mm.
Contact pressure	12 oz.
Condenser capacity	·29 mfd.
Condenser internal resistance	·21 ohms
Contact closed period 55° with ·012 in. gap.	

Ignition Coil, Type TL6

The coil is of the orthodox outside primary type assembled into a can on which the fixing strap is secured by spot-welding. The strap is provided with two elongated holes for securing to the engine and the insert in the moulded top is rolled over the edge of the can. The primary leads are brought through and soldered to 2 B.A. terminals moulded in the coil top, and the high tension terminal is of the push-in type. The seams in the can do not appear to be soldered.

The resistance of the secondary winding is 3,700 ohms, and the primary inductance 9 M.H. The primary standstill current is of the order of 5 amperes and the temperature rise 119° C. By way of comparison, the Lucas 6Q6 coil has a standstill current of approximately 4 amperes and a temperature rise of 100° C.

Dimensions of the coil are as follows:

Diameter of can	2 $\frac{3}{64}$ ins.
Length of can	4 $\frac{7}{16}$ ins.
Overall height (approximate, owing to broken moulding)	5 $\frac{3}{4}$ ins.

H.T. Cables and Suppressors

The cable is of the 7 mm. rubber covered type having an unusually large core of 64 strands of ·012 in. tinned copper wire (as compared with 40 strands of ·010 in. wire used in standard English 7 mm. cable and 19 strands of ·012 in. wire in the latest screened cable). As a result of this large core the rubber section of the cable is considerably reduced, there being approximately only 2 mm. of rubber per side. This considerably reduces the dielectric strength. The times of breakdown with different potentials are as follows:

12 KV	6 min. 0 sec.
15 KV	2 min. 50 sec.
20 KV	30 sec.

Standard Lucas 7 mm. rubber covered cable withstands 20 KV for five minutes. Trouble in service may therefore be expected with the German cable.

The rubber insulation, which is made up with an inner layer of white and an outer layer of grey rubber, is very loose on the cable core. The formation of air pockets adjacent to the core is very undesirable, as they are liable to produce premature ozone failure of the inner rubber dielectric. Ozone test results suggest that the rubber dielectric is not synthetic, failure occurring after 1 min. 30 sec. at 20 KV potential.

At the plug end of each lead is a bakelite moulding which houses the suppressor. The bakelite is a wood-filled material, and the suppressor is wire wound and connected to brass end caps. Contact with the cable is made by a wood screw secured in the moulding and the suppressor is secured between the screw and a plug screwed into the other end of the moulding. A spring clip in the screwed plug enables it to be clipped on to the sparking plug. A flange-shaped rubber is threaded over the suppressor moulding, presumably to prevent the sparking plug pockets in the cylinders from becoming filled with dirt and sand. The suppressor in the coil lead is of the usual tubular type with wood screws at each end.

Details of the suppressor are as follows:—

Overall length of moulding	2 3/8 ins.
Length of suppressor resistance	·900 in.
Suppressor resistance	10,000 ohms
Flange diameter of the rubber	1 3/4 ins.
Length of tubular suppressor	2 3/16 ins.

BOSCH DYNAMO,
TYPE RED 130-6 2600 AL89

The dynamo is of the six-volt type and arranged for negative earth connection. Rotation is anti-clockwise as viewed from the driving end.

Plate 42.—Dynamo body, showing field winding, brush holder and regulator

The dynamo is a 3½ ins. outside diameter ventilated machine, with an F Type regulator unit mounted on the yoke. Extended through bolts are provided at the driving end, so that the machine can be mounted on a flange incorporated in the engine casting. The shaft is extended at both ends to provide for a belt drive at the driving end, and it is assumed that a fan is fitted at the other end, although this was not on the dynamo when received.

The field coils are taped with crepe paper and treated with the usual varnishing process. The regulator is a Bosch F Type, in which both the cut-out and regulator are operated by one armature. A diagram of the connections and terminal markings is shown below. The contacts were badly burnt and adjustment had to be made to the second contacts on the regulator before the latter would operate.

In other respects the dynamo and regulator units follow conventional design.

Curve No. A/2439 shows the output of the dynamo under hot and cold conditions, whilst Curve No. 2440 shows the temperature rise on the yoke after a heat run of 2 hours at 6·75 volts, 20 amperes and 3,000 r.p.m. A pulley incorporating a fan was fitted to the dynamo for the purpose of the heat run.

Weight of dynamo, complete with regulator	11¼ lb.
Open circuit regulated voltage	7·0 volts
Cut-in voltage of cut-out	7·4 volts
Drop-out voltage of cut-out	4·8 volts
Regulator points resistance	7·5 ohms

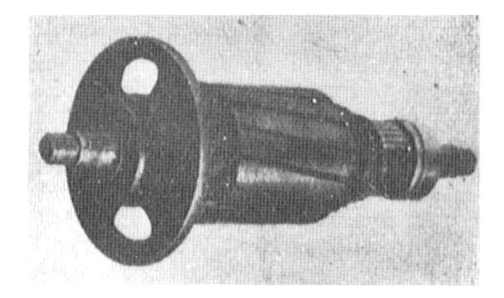

Plate 43.—Dynamo armature and commutator assembly

Design Data

Yoke: Outside diameter 3½ ins.; inside diameter 3 1/16 ins.; length 5 13/16 ins.

Field: Number of poles 2; pole chord 1·969; pole arc/pole pitch 0·60; bore of poles at centre 2·33 ins.—at tips 2·33 ins.; air gap length ·055 in. per side.

Armature: Diameter 2·22 in.; length 2·5 in.; Number of slots 15; type of slot—open, skewed.

Commutator: Diameter 1·272 ins.; length 15/16 in.; number of segments 30.

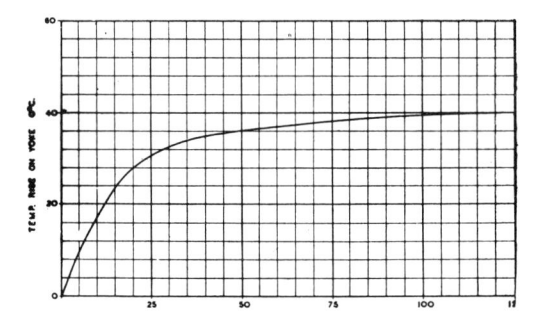

Fig. 44.—Distributor, temperature rise

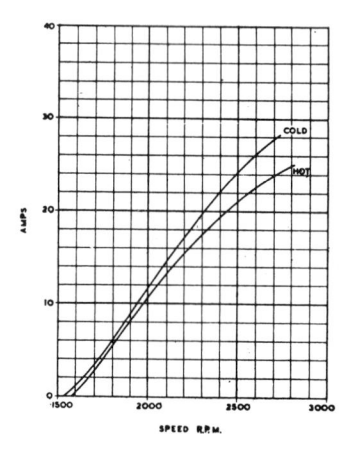

Fig. 45.—Distributor, output

55

Brushes: Type of support—box; trigger spring; brush arms 2; brushes per arm 1; length $\frac{17}{32}$ in.; width $\frac{7}{32}$ in.; copper type.

Bearings: Commutator end — diameter 1·378 ins.; width $\frac{7}{32}$ in.; ball bearing. Driving end—diameter 1·378 ins.; width $\frac{7}{32}$ in.; ball bearing.

BOSCH STARTER, TYPE EEDD 4-6 L3P

The starter is of the six-volt type and arranged for negative earth connection. Rotation is anti-clockwise as viewed from the drive end. The starter has an outboard drive of the "screw-push" type with solenoid engagement, the solenoid being mounted on an extension integral with the drive end bracket. The commutator end bracket is secured to the yoke by means of four projection welds, the drive end bracket being secured by

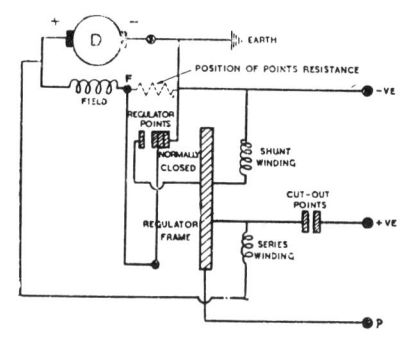

Fig. 46.—Distributor, circuit

means of two hook bolts. These methods of securing the end brackets are employed because, as the result of fullest use having been made of the available field winding space, there is no room for through bolts.

There is no intermediate bracket or bearing, the second bearing being provided in the flywheel casing when the starter is fitted to the engine.

The pinion is provided with a free wheel to prevent it becoming damaged when over-run by the engine flywheel.

An interesting feature of the starter is a device for slowing down the armature when the pinion is disengaged, so that if a second engagement is necessary the pinion will have ceased to rotate or will be moving only slowly when re-engagement is made. This slowing down is obtained by means of a small friction clutch operating at the commutator end in the following manner:

Plate 44.—Starter body, showing field windings and brush holder

A plate is fitted on the shaft near the commutator, two flats on the shaft ensuring drive for the plate. On the plate are two dogs, diametrically opposite, fitting into two semi-circular brake shoes located in a housing adjacent to the bearing. Loading is applied to these two shoes by means of a jump ring located in a groove in the brake shoes. The approximate torque required to turn the armature against this clutch is 1·6 lb. inches.
Weight of starter (with solenoid): 13¼ lb.
Weight of starter (without solenoid): 12¼ lb.

Plate 45.—Starter armature and commutator

Design Data

Yoke: Outside diameter $3\frac{9}{16}$ ins.; inside diameter $3\frac{1}{8}$ ins.; length $5\frac{1}{2}$ ins.

Field: Number of poles 4; pole chord 1·25 ins.; pole arc/pole pitch ·67; bore of poles at centre 2·403 ins.—at tips 2·403 ins.; air gap length ·020 in. per side.

Armature: Diameter 2·363 ins.; length $2\frac{7}{16}$ ins.; number of slots 23; type of slot—semi-enclosed, straight.

Commutator: Diameter 1·409 ins.; length $\frac{7}{8}$ in.; number of segments 23.

Brushes: Type of support—box; spring lever. Two brushes—length $\frac{5}{8}$ in.; width $\frac{9}{32}$ in.; depth $\frac{3}{4}$ in.; grade 131.

Pinion: 9 teeth.

SOLENOID STARTER SWITCH

The solenoid starter switch is mounted on an extension integral with the drive end bracket of the starter.

It consists of a solenoid of conventional design with a conical plunger and core. A push rod is fitted at the conical end of the plunger which carries a copper contact plate. When the switch is operated the spring-loaded contact moves to bridge two contacts fitted in the end of the switch housing, and so completes the circuit from the battery to the starter. At the opposite end of the plunger is fitted a forked extension piece which is connected mechanically to the starter pinion. Thus operation of the starter switch—in addition to completing the circuit to the starter—also causes the pinion to move into engagement with engine flywheel. A spring is provided which ensures that when the switch is released the contacts are separated and the pinion is moved out of engagement with the flywheel.

The solenoid is provided with two windings, one of high resistance and the other low. The reason for this is not known, but it may be necessary to obtain satisfactory engagement of the pinion.

Plate 46.—Solenoid exterior, with cover removed

Design Details

Resistance between main contacts and winding terminal = 0·14 ohms.
Resistance between main contacts and case = 0·32 ohms.
Pull-on volts = 4·1 with low resistance winding.
Drop-off volts = 0·4.
Pressure on plunger to close contacts = $9\frac{1}{4}$ lb.
Pressure on plunger to push right home = $19\frac{1}{4}$ lb.
Volt drop across contacts with 6 volts on low resistance winding = ·059 at 20 amps.
Plunger travel to close contacts = $\frac{5}{16}$ in.
Total plunger travel = $\frac{13}{32}$ in.
Weight of unit = $1\frac{1}{4}$ lb.

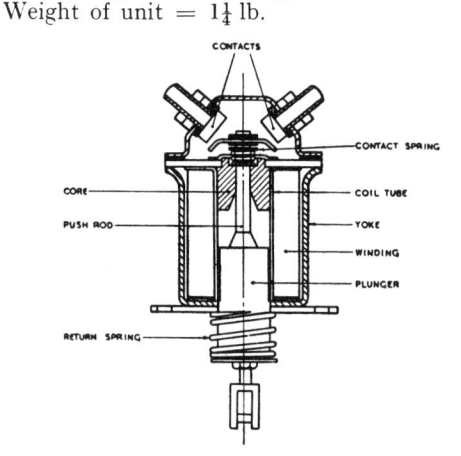

Fig. 47.—Solenoid, section

Dimensions :
Length of yoke = $1\frac{7}{8}$ ins.
O/D of yoke = $1\frac{7}{8}$ ins.
I/D of yoke = $1\frac{3}{4}$ ins.
Plunger diameter = 0·748 in.
Plunger angle = 60°.

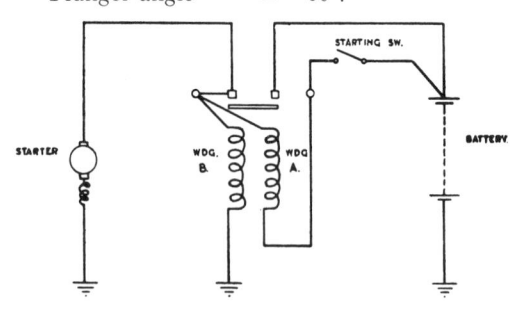

WINDING A : RES.=0·14" M.O.V.=4·1 D.O.V.=0·4
WINDING B : RES.=0·32" M.O.V.=10·4 D.O.V.=0·9

Fig. 48.—Solenoid, connections

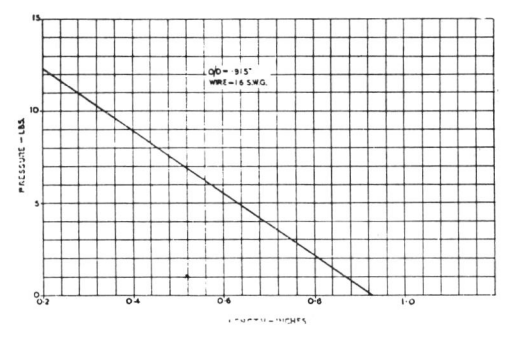

Length of core = $\frac{7}{32}$ in.
Length of plunger = $1\frac{5}{8}$ ins.
Length of core = $\frac{5}{8}$ in.
O/D of winding = $1\frac{5}{8}$ ins.
I/D of winding = $\frac{25}{32}$ in.
Length of winding = $1\frac{5}{8}$ ins.
S.W.G. of wire = 19.
Push rod diameter = 0·236 in.

Fig. 49.—Solenoid, characteristic of plunger return spring

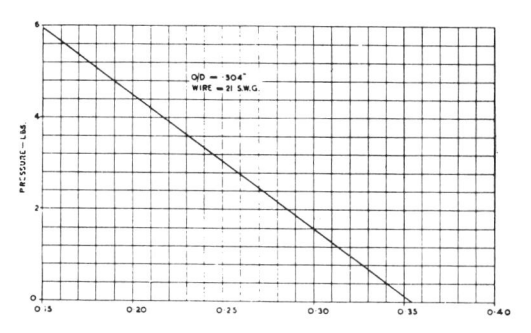

Fig. 50.—Solenoid, characteristic of contact spring

VARTA BATTERY, TYPE 937B

The battery is a six-volt unit of conventional design as shown below.

Plate 47.—Battery, showing cell removed

Constructional Details

Weight with acid : 36 lb. 2 oz.
Dimensions : Length over lugs, $8\frac{7}{8}$ ins. ; length without lugs, $8\frac{7}{16}$ ins. ; width, $6\frac{7}{8}$ ins. ; height over terminals, $8\frac{1}{2}$ ins. ; height over container, $7\frac{9}{16}$ ins.
Container and lid material : Low grade ebonite.
Plates : 13 per cell. Dimensions : $5\frac{5}{8}$ ins. × $4\frac{7}{8}$ ins. ; positives, $\frac{3}{32}$ in. thick ; negatives, $\frac{1}{16}$ in. thick.

Separators: Wood, grooved on one side with thick webs.

Vent plugs: Porcelain, centre hole and screw thread type.

Baffle plate or separator protector fitted.

Performance

Condition as received: Partially discharged, specific gravities of 1·194, 1·205 and 1·168 (at 60° F.) per cell respectively.

Charged at 6·5 amps. for a total period of $28\frac{3}{4}$ hours.

Maximum temperature during charging 95° F.

Specific gravities of three cells after conditioning charge: 1·251, 1·262 and 1·258 at 60° F.

First Discharge at 6.3 amps. without adjusting specific gravity.

Capacity 14·2 amp. hours to 5·40 volts at 71° F.

Cell 1.   22·0 amp. hours.
Cell 2.   12·3 amp. hours.
Cell 3.   14·5 amp. hours.

Second Discharge at 6.3 amps. after adjusting electrolyte to 1·292 specific gravity at 60° F.

Capacity 14·5 amp. hours to 5·4 volts at 80° F.

Cell 1.   28·5 amp. hours.
Cell 2.   15·0 amp. hours.
Cell 3.   16·5 amp. hours.

Continuous Discharge at 189 amps. commencing at 75° F.

Specific gravity before discharge: 1·295 (at 60° F.).

Voltage after 5 secs.: 3·80.

Duration of discharge to 3·0 volts: 42 secs.

Owing to the poor performance of the battery at 75° F. no further tests were made at 0° F. or —21° F.

SWITCHGEAR AND INSTRUMENTS

The speedometer, lighting switch, speedometer light switch, ignition switch and warning lights are fitted in a moulded panel which is mounted on a metal facia plate incorporating the starter operating switch, control switch for hooded headlamp and distance indicating lamp, inspection lamp socket and two fuse boxes.

Main Lighting Switch

This is fitted on the right-hand side of the moulded panel. It controls the main and pilot bulbs of the two normal headlamps and also the tail lamp. An interesting feature of the switch is that the stationary contacts are moulded integrally with the panel plate, whilst the rotor assembly is incorporated in the circular control knob. The switch has three positions with a central "off" position.

Speedometer Light Switch

The design of this switch is similar to that of the main lighting switch. It is fitted on the left-hand side of the moulded panel.

Ignition Switch

This is fitted at the bottom of the moulded panel. It is of the normal Yale lock type of construction. Owing to the switch having been damaged it was not possible to carry out any tests.

Switch controlling Hooded Headlamp and Distance Indicating Lamp

This switch is mounted on the right-hand side of the steel facia plate. It incorporates a wire-wound resistance which is arranged to be connected in series with the hooded headlamp so that the light given by the lamp can be adjusted according to operational conditions. The switch has five positions as follows:

O. Hooded headlamp off. Distance indicator lamp off.

H. Hooded headlamp off. Distance indicator lamp on.

V. Hooded headlamp connected through 0·95 ohms resistance. Distance indicator lamp on.

V. Hooded headlamp connected through 0·4 ohms resistance. Distance indicator lamp on.

V. Hooded headlamp fully on. Distance indicator lamp on.

The resistance is wound round the ribbed body of the switch, which is approximately $2\frac{1}{8}$ ins. diameter and $1\frac{3}{4}$ ins. long. The resistance wire is protected by a perforated steel outer cover and has a diameter of ·052 in. and a length of approximately 4 ft.

Starter Operating Switch

The original switch has been removed and replacement Lucas unit fitted.

Dipper Switch

The dipper switch controlling the change-over from the main driving light filament of the headlamps to the dipped filament is fitted in the floor board on the left-hand side. It is of the standard "push-push" design arranged for foot operation.

Inspection Lamp Socket

This is of the single pole "push-in" type.

Plate 48.—Interior, showing facia panel

Warning Lamps

Four warning lamps are provided in the moulded instrument panel, two being fitted on either side of the speedometer. Apparently only two of the warning lamps are used, the top one on the left-hand side (coloured red), for the ignition, and the bottom right-hand (coloured blue), for the head or spot lamps.

Plate 50.—Facia panel, rear view, showing wiring

Plate 49.—Facia panel, front view, showing fuse box cover removed

The warning lamp bulbs are 6 volt, 1·2 watt; the approximate dimensions of the glass bulb being ·270 in. diameter and ·25 in. long. The bulbs are provided with miniature bayonet caps of the centre contact single pole type. The bulbs are housed in tubular brass spinnings ·378 in. diameter and $\frac{23}{32}$ in. long. This assembly—including a spring loaded contact plunger—fits into a raised portion at the back of the moulded panel.

Fuses

Two fuse boxes are fitted on the metal facia plate. Each box houses five fuses protected by a moulded cover, on the inside of which an illustrated card is fitted to identify the circuits which the fuses protect. The fuses are of the 15 amp. size and have steatite bodies with metal end caps. The tinned copper fuse wire is located in a groove in the steatite body and the ends are secured under the metal caps.

The fuses are secured in position by locating one end in a conical recess in a terminal while the other end fits in a location in a spring contact blade.

These fuses are connected in the following circuits :—

Left-hand Fuse Box
Electric Horn.
Pilot bulbs in left- and right-hand headlamps.
Dipped beam filaments in left- and right-hand headlamps.
Main filament in left-hand headlamp.
Main filament in right-hand headlamp.

Right-hand Fuse Box
Hooded driving lamp.
Distance indicating lamp.
Trafficators and windscreen wipers.
Spot light.
Inspection lamp.

In addition, a separate fuse box containing three fuses is fitted in the engine compartment at the rear of the vehicle; these fuses are provided for the protection of the left-hand tail lamp, right-hand tail lamp and the stop lamp.

A test was carried out on one of the fuses, which blew at 25 amps. when the current was raised at the rate of 2 amps. per second. The total voltage drop across each pair of terminals with 10 amps. passing was ·240—·260 volts.

WIRING

The wiring incorporates numerous types of cables. Some consist of bare copper wire with synthetic covering, some are of tinned copper with a cotton covering and a cotton

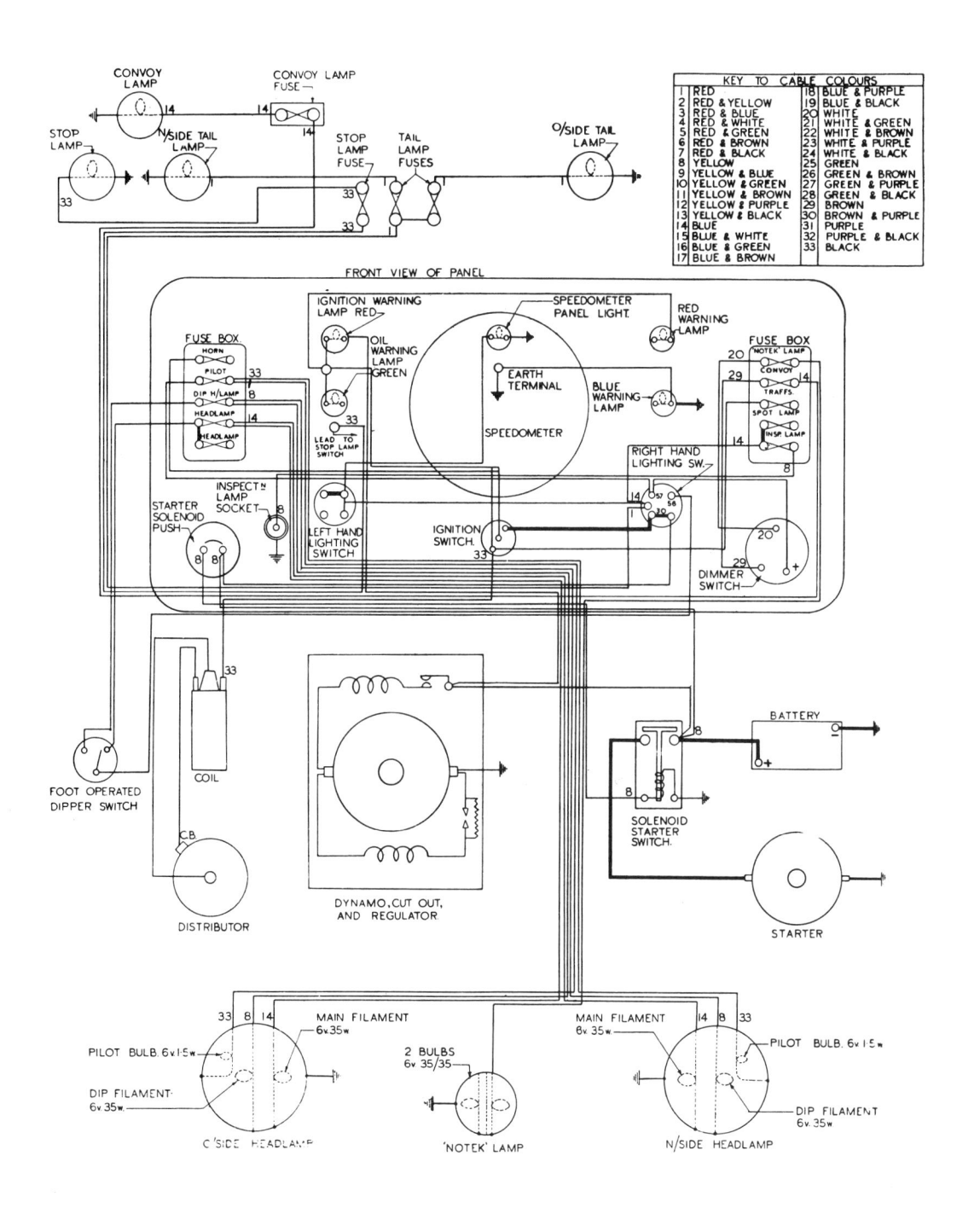

Fig. 51.—Wiring diagram

braiding overall with no rubber insulation, whilst still more are of tinned copper wire with a wrapping ·001 in. thick of black varnished material and varnished cotton braiding. The starter cable is 4 ft. 6 ins. long and is of ·442 over-all diameter. It consists of 37 strands of ·044 in. diameter conductors with a covering of rubber ·060 in. thick and black cotton braiding.

A 19/·024 cable is used for the main dynamo circuit. It has a polyvinyl chloride covering of ·025 in. thickness, the over-all diameter of the cable being ·168 in.

The negative terminal of the battery is earthed to the chassis by a bare copper braid 11 in. long, 1¼ ins. wide, and ·080 ins. thick.

LIGHTING EQUIPMENT

The lighting equipment comprises the following components:

(a) 2 Headlamps of identical type. On one lamp the aperture was restricted by the fitting of a leather cover and on the other by means of paint.
(b) 1 Notek Headlamp provided with single-slot mask and cowl.
(c) 1 Hella spot light.
(d) 1 Notek distance indicating rear lamp and stop light.

(a) Headlamps

Each lamp contains an aluminium reflector held in place by a single spring. The reflector is burnished and lacquered on the inside, but has a poor finish. Both main and pilot bulbs are mounted on a plate at the back of the reflector in a similar fashion to Lucas D Type lamps. The main bulb is of Lucas-Graves type with 6 volt 35 watt filaments. The pilot bulb is a single pole 6 volt 3 watt midget type. Contact is made by brass clips to which the connecting wires are attached.

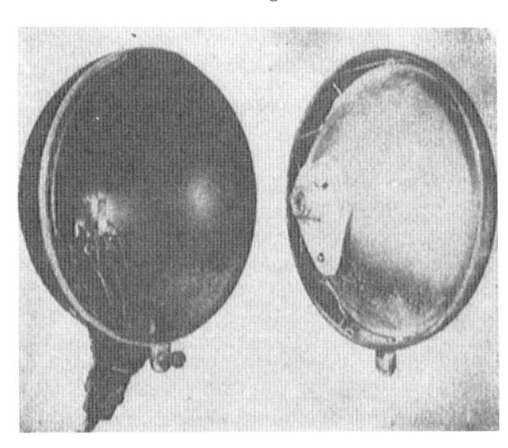

Plate 51.—Headlamp

The front glass has a total diameter of 17 cms. (6·8 in.). The effective aperture, however, is a central spot of about 4×2 cm. ($1·6 \times 0·8$ in.). There is no positive location of the rim, which is held in position by a single screw. The maximum output is 475 candles with the main filament in action, and 300 candles with the dip filament. The total spread of the beam is 15° in the horizontal and 4° in the vertical plane.

(b) Notek Headlamp

The lamp body consists of two parts connected by three screws. The bottom piece is die cast, carrying the reflector and bulb, and the top, containing the lens and cowl, is of pressed steel. Between the two is fitted a rubber gasket. The bulb is of Lucas-Graves type, using only the Vee filament, which is rated at 6 volts 35 watts. It is horizontally mounted with the filament towards the reflector. The bulb has a locating device and can only be removed by releasing two screws.

There is no wiring in the lamp, connection being made by a plug socket fitted with a locking device.

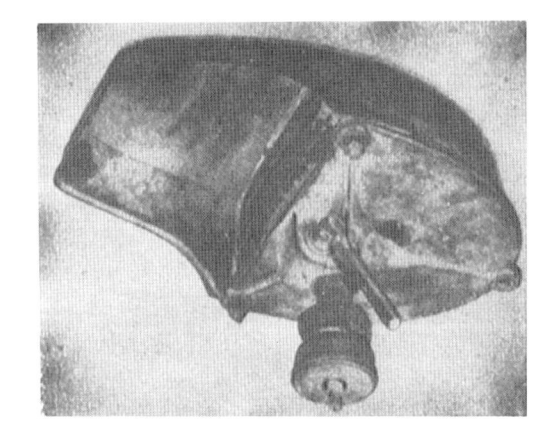

Plate 52.—Notek headlamp

The reflector has an effective width of 2·25 cm. (0·9 in.) and a length of 13·5 cm. (5½ ins.) at the aperture. The focal length is 2 cm. (0·8 in.). The mirror body forms a frame for the purpose of preventing distortion, the axis of the mirror being dipped slightly below the horizontal.

The front glass consists of plano-convex flutes, five flutes per inch; it is held in

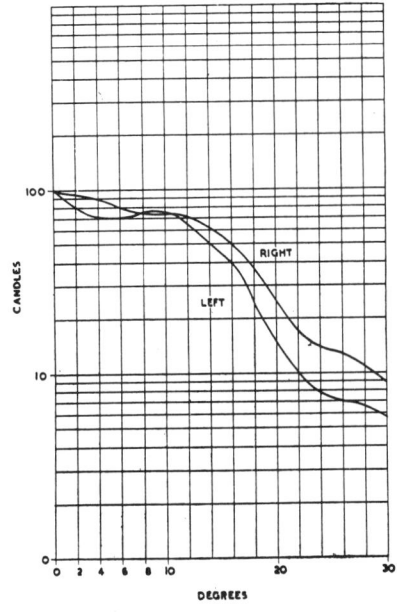

Fig. 52.—Notek headlamp, horizontal beam distribution

place by a spring device. The effective aperture is 1·2 × 10·5 cm. (0·48 in. × 4·2 ins.). The length of the cowl is 10·5 cm. (4·2 ins.) (measured from the front of the lens), and cuts off 7 mm. below the bottom edge of the lens aperture.

The beam distribution curves shown in Figs. 52 and 53 were taken with the lamp as received, with dirty reflector and lens. The maximum brightness of the beam was 1·0 F.C. After cleaning, this rose to 1·4 F.C. The cut-off is very sharp and 6° below the horizontal plane.

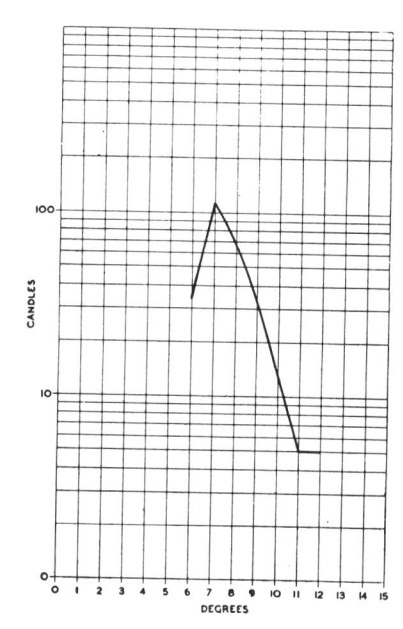

Fig. 53.—Notek headlamp, vertical beam distribution

(c) Hella Spot Light

The lamp body functions as the reflector; it is of brass, painted on the outside and silvered and lacquered on the inside. The front glass is fixed to the reflector by a split ring which is tightened by a screw to hold the glass in position. The glass is marked "Hella" and is clear in the centre with three flutes on either side. The diameter of the glass is 11 cm. (4·4 ins.). The focal length of the reflector is 2·5 cm. (1 in.).

The bulb was broken when received, consequently no photometric measurements have been carried out. It is rated at 6 volts 25 watts

Plate 53.—Hella spot light

with a coiled coil hacksaw filament and single pole adaptor. At the back of the lamp is a switch acting on a contact spring on the bulb socket. Over the whole lamp there is a rubberised canvas cover with a narrow slot in the centre 2·5 × 0·5 cm. (1 in. × ¼ in.).

(d) Notek Distance Indicating Rear Lamp and Stop Light

The die-cast body contains three 6 volt 10 watt festoon bulbs, two mounted horizontally and one vertically. The vertical bulb illuminates the distance indicator lamp. The distance indicator has four rectangular windows, the two outside windows being larger (2·2 × 3·2 cm.) than the two inner windows (1·8 × 3·0 cm.). The distance between the two inner windows is 3·8 cm., and between the two pairs of windows on either side of the centre it is 1·5 cm.

The underlying principle of the distance indicator is the dependence of the resolving power of the eye on the apparent angle of an object.

Plate 54.—Notek distance indicator, rear lamp and stop light

The four windows indicate two limit distances.
 1. The fusion of the two pairs of windows.
 2. The fusion of all four windows.

Taking the average angular distance for the resolving power of one degree, these two distance limits will be 50 metres and 140 metres respectively. The difference in the sizes of the outer and inner windows enables the observer to estimate intermediate distances according to the extent to which the smaller window merges with the larger.

The stop-tail lamp and rear lamps make use of the two horizontally mounted festoon bulbs and have orange and red glasses respectively. The change-over from normal running to convoy work is effected by a hinged flap held in position by spring clips. This flap contains a hole 0·5 mm. diameter which is positioned over the rear lamps when the flap is in the convoy position.

A window (which can be obscured by a shutter) illuminates the number plate of the vehicle.

ELECTRIC HORN manufactured by the Metallische Industrie A.-G., Lippstadt

The horn is of the H.F. type, i.e., an armature attached to the diaphragm impacts on the magnet core and at the same time operates a side contact breaker. To make the side thrust of the contact breaker as small as possible, the magnet core is very narrow and rather long, and is of laminated construction. This is offset to some extent by the abnormally high contact pressure : 7 lb. instead of the usual 3-4 lb. It is not obvious why this is so. Adjustment of the contact breaker from the back of the horn is by means of a spring-loaded screw which raises the contact breaker bodily about the other fixed end.

Plate 55.—Electric horn, exterior

The high frequency component of the note is generated by a " swinging plate " attached centrally to the underside of the diaphragm, between the latter and the laminated armature —this, presumably, because the tone disc is still a Bosch patent. It is not nearly so effective as the external tone disc.
The special feature of this horn is that the body and front grille are in moulded bakelite instead of metal pressings or castings. The magnet core is moulded in the body, a bakelite flash round the core providing the coil insulation. Plain enamelled wire is used for the coil, which is retained in the core by two clips.
The contact breaker bracket also carries a small cylindrical condenser, attached by a single through bolt.

Plate 56.—Electric horn, interior

The moulded construction is very neat and light and allows the terminals to be incorporated directly without any additional insulation. The front grille has holes tapped directly in the bakelite (moulded threads). The average thickness of the bakelite body wall is only ·10 ins. to ·125 in.
A bolt attached to the magnet core projects through the back of the body and carries a laminated spring bracket of Bosch pattern.

Performance

The sample examined was a 6 volt horn.

Current Consumption

	German horn	HF.1235
4 v.	2·0	
6 v.	3·1	3·0—4·0
7·6 v.	3·8	4·0—5·0

The current consumption is fairly low, some advantage being gained from the use of a laminated magnet.

NOTE.—The fundamental frequency is approximately that of the Lucas Type HF.1235. The note is fairly loud and clear over the voltage range, but the tone is rather flat with a less well-marked high frequency whistle than that of the Lucas horn, the German horn having a fairly distributed frequency spectrum. Its overall loudness is slightly greater than that of the Lucas horn due to the larger diaphragm.

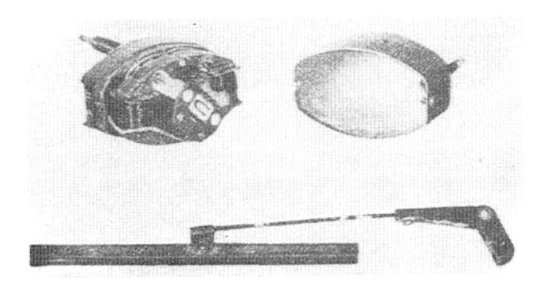

Plate 57.—Windscreen wiper, exploded view

General Remarks

The diaphragm was very rusty, the paint protection being inadequate. The front grille was broken in two places and it is suggested that although the bakelite construction has many advantages it is not sufficiently robust for Army vehicles.

WINDSCREEN WIPER

Two separate windscreen wipers are provided, and these are fitted at the bottom of the screen—on the off and near sides respectively.

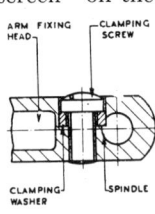

Fig. 54.—Wiper arm and spindle section

There is no marking on the wiper to indicate the make but it is similar in general shape, motor and gearing to the Bosch W.V. Type wiper. The wiper is marked " 12V " on its back cover, but all the electrical equipment on the vehicle is 6 volt. From its performance on a 6 volt supply it would appear that the marking on the wiper is incorrect.
Details of the design and performance of the wiper are as follows :

Approximate External Dimensions

Length : 3 3/16 ins.
Maximum width : 1 15/16 ins.
Minimum width : 1 in.
Overall height (excluding wiping spindle bearing but including bearing flange) : 2 1/2 ins.

Plate 58.—Windscreen wiper, rear cover removed

Wiping spindle bearing : $1\frac{3}{8}$ ins.
Spindle extension through bearing : $\frac{5}{8}$ in.
Length of switch lever : $1\frac{1}{32}$ ins.

Total Weight
18 oz. approximately.

Details of Field System
Number of poles : 2.
Inside diameter between poles 1·420 ins. and 1·432 ins. (difference due to incorrect setting of pole halves).
Height of laminations : 10·5 mm.—11·0 mm.
Pole arc : 120° approximately. Arc length 37 mm.
Area of pole face : 400 sq. mm.
Field shunt connected, consisting of 2 coils in series having a total resistance of 6·2 ohms.

Details of Armature
Number of poles : 3.
Outside diameter : 1·390 ins.
Air gap (average) : ·018 in. per side.
Armature core length ·473 in. (12 mm.).
Pole arc 80° approx. Arc length 23 mm. approx.
Resistance between segments : 1·5 ohms.

Commutator
Number of segments : 3.
Overall length : approximately $\frac{5}{16}$ in.
Diameter on brush track : 0·393 in.

Brushes
Dimensions : $\frac{5}{16}$ in. × $\frac{3}{16}$ in. × $\frac{1}{8}$ in. approx.
Not fitted with pigtails.
Spring pressures : 110 gms. each.

Circuit
Earth return wiper. The switch lever completes the circuit from a brass strip fastened to the brush plate (and connected to one brush box), to an earthed terminal via the cover.

Gearing and Final Drive Details
Armature pinion : 6 teeth.
Inter-gear ("Tufnol" or similar material) : 39 teeth.
Inter-pinion : 7 teeth.
Guide wheel : 70 teeth.
Total gear ratio : 65 : 1 reduction.
Oscillating mechanism consists of a rack pivoted at a radius of $\frac{5}{16}$ in. on the guide wheel, driving a toothed wheel of approximately $\frac{1}{2}$ in. pitch circle diameter and mounted on the final wiping spindle, thus giving a theoretical angle of wipe of 140°.
The following performance test results were obtained when driving the single arm and blade (details of which follow) on a dry clean screen. Wiper at room temperature.
Applied voltage : 6 volt.
Current : 1·9—2·0 amps. (approximately).
Wiping speed : $26\frac{1}{2}$ wipes per minute.
Angle of wipe : 125°.

After Half-hour Run
Current : 1·4—1·5 amps. (approximately).
Wiping speed : 35 wipes per minute.
Heating : satisfactory.

Further Tests taken at Room Temperature
Oscillating shaft stall torque : 150 oz.-in. (approx.)
Armature stall torque : 6 oz.-in. (approx.).
Stall current : 4·1—5·0 amps. (approx.).
Armature speed (running light) : 1,750 r.p.m.
Light running current : 1·98 amps.
Armature flux with 6 volts on field (cold) : $11\frac{1}{4}$ K.L.

Arm and Blade Details
Effective length of arm (from spindle centre to blade pivot point) : $7\frac{1}{4}$ in. (approx.).
Length of blade : 7 ins.
Blade pivot point from end furthest from driving spindle : $3\frac{3}{4}$ in.
Blade consists of 2 plys of rubber, ·020 in. thick, with a metal separating piece inside the channel.
The fixing means for the arm consists of a clamping washer which is forced against the arm by pressure applied through the clamping screw.

ACKNOWLEDGMENTS

In compiling this Report we desire to acknowledge the valuable assistance given by the following firms in respect to the information concerning the items stated :

Messrs. JOSEPH LUCAS LTD.
  Electrical Equipment and Shock Absorbers.

Messrs. PRESSED STEEL CO., LTD.
  Body and Underframe.

Messrs. DUNLOP RUBBER CO., LTD.
  Wheels and Tyres.

Engineering Dept.
HUMBER LTD.
COVENTRY

Sept. 1943.

PART 11.

COMPARATIVE ROAD PERFORMANCE TEST OF THE
VOLKSWAGEN SALOON TYPE 11.

Report by:- HUMBER LTD.

A civilian saloon version of the above car built in Germany under British jurisdiction since the conclusion of hostilities, was received by the Humber Experimental Department on the 18.2.46. Tests have been carried out for the purpose of obtaining performance data and general impressions of the vehicle, but it should be emphasised at the start that no dismantling was carried out, and therefore the description of the vehicle which is given covers only such information as could be obtained from a superficial examination.

Where the comparative performance figures are put alongside those of a Hillman Minx, the Experimental Minx M.3 was used, and no special tuning was carried out before the figures were taken, it being considered fairer to pick a car at random in average condition for comparative purposes.

For the benefit of those not familiar with the test hills referred to in Section 4 of the comparative test data, the following description is given.

Frizz Hill.

Average gradient 1 in 23.5. Approx. length 400 yards. Straight hill climb gradient, easing towards the top.

Edge Hill.

Average gradient 1 in 8. Steepest 1 in 7.1. Approx. length 800 yards. One sharp R.H. bend near the bottom with gradual left and right hand bends near the top. Steepest portion approx. 50 yards after R.H. bend.

- 65 -

ENGINE.

The engine on this vehicle was almost identical with that of the previous military version tested, but we understand that the cubic capacity has been increased from 960 c.c.'s of the earlier model to 1161 c.c.'s for this civilian saloon model. The engine is of course of the horizontally opposed 4 cylinder air cooled type, employing push rod operated overhead valves with a forced draught air cooling employing a blower mounted on the top of the engine. The engine is mounted in the rear of the vehicle, forming one unit with the clutch, gearbox and final drive, and is carried on three fixing points with no visible rubber insulation. The mounting points are located at the bell housing and at the nose end of the gearbox.

The drive is understood to be taken through the normal dry plate clutch, which is operated by means of a cable from the driving position, which is on the left hand side. The clutch pedal pressure is 40 lbs. and the travel $4\frac{3}{4}"$.

The engine is somewhat noisy and rough and no doubt the absence of water jacketing and soft mounting are responsible for this. As will be seen from the performance and petrol consumption figures quoted later in this report, the performance is by no means outstanding but the consumption figure is quite good, although of course this is partly due to the extreme lightness of the vehicle. The size of the carburetter and induction pipe suggests that more power could be obtained from the engine as these parts appear to have been reduced deliberately in size, but whether the engine is capable of withstanding an increased power output is a debatable point.

As a check on the engine condition before the tests were made, the compression pressures of the 4 cylinders were checked and the following readings were obtained.

N.S.F. cylinder	113 lbs./sq.in.
O.S.F. "	86 "
N.S.R. "	115 "
O.S.R. "	106 "

For comparative purposes the following figures were obtained from the Experimental Minx M.3, which was used.

No.1 cylinder	125 lbs./sq.in.
No.2 "	121 "
No.3 "	123 "
No.4 "	121 "

It will be seen that one of the cylinders in the case of the Volkswagen was rather badly down, but no action was taken to dismantle the engine to find the cause. In the case of the Minx, it will be seen that the general run of compression pressures was higher and they were satisfactorily even, indicating that the valves, piston rings etc., were in good condition. Due allowance should be made therefore when assessing the comparative performance figures for the unknown condition of the engine of the Volkswagen.

In spite of this however, it is considered that the performance should have been better in view of the lighter weight of the Volkswagen. When both cars were weighted similarly, i.e. Minx 3 up = Volkswagen 5 up + 160 lbs, this difference in performance is very marked. It is interesting also to compare the performance of each car in its fully laden condition, i.e. Minx 4 up + 200 lbs and Volkswagen 5 up + 160 lbs. Reference should be made to Section 4 of the comparative test data.

GEARBOX.

Four speed and reverse gearbox is supplied, none of the gears having synchromesh operation, but the gear change is a particularly easy one. No examination of the gearbox was made, but the military version which was examined some time ago had an ingenious type of dog clutch engagement for Top and 3rd gears, and the ease with which the gear change can be made suggests that the arrangement in the civilian version is similar. The gears are operated by means of a remote control lever conveniently mounted between the two front seats, the operation of which is clean, light and positive and entirely free from engine or road interference. First and second gears are somewhat noisy, but the remainder are satisfactorily quiet and appear to be suitable for the purpose for which they were designed. From the gearbox and rear axle unit, the drive is taken through metal universal joints to the half shafts, which are encased in tubes, to whose outer ends are bolted the brake backing plates and hub bearings. The universal joints are covered by flexible rubber gaiters and no hub reduction gear is fitted.

STEERING.

The steering is of the orthodox type, assumed from external appearance to be worm and nut or worm and wheel, giving a rather direct feel and taking a good deal of getting used to. There are just over $2\frac{3}{4}$ turns of the steering wheel from full lock to full lock, and details of the turning circle are given in a separate section under comparative test data. The steering is distinctly tricky on bends, and has a tendency to over steer. The drop arm is roughly 'T' shaped, carrying a two piece track rod very similar to the standard Humber I.F.S. arrangement. A fabric universal joint has been introduced in the steering column and the steering was noted to be particularly free from kick.

SUSPENSION.

Suspension front and rear is by means of torsion bars anchored centrally to the main backbone of the chassis. The action of the torsion bar is controlled by hydraulic shock absorbers, and whilst it may have been reasonably suitable for a military or cross country vehicle, the suspension in our opinion is of a very poor standard for a civilian vehicle.

The suspension is hard and choppy and over rough or smooth surfaces gives a very poor account of itself, giving the occupants a very poor ride.

The shock absorbers are of the piston type at the rear and of the telescopic direct acting type on the front.

BRAKES.

The brakes are of the cable operated type on all four wheels. The hand brake, the operating lever of which is placed centrally between the seats, operates via the same cables on all wheels also. Brake performance is definitely poor, as will be seen from the figures given in the comparative test data.

ELECTRICAL EQUIPMENT.

Flush fitting combined head and side lamps are fitted and a stop/tail lamp mounted centrally on the engine cover. The headlamps are of the double filament type, controlled by a foot operated dipper switch. Dual electric non parking screenwipers are fitted, controlled by a two-way switch which also operates the interior light. No indicators are fitted and the electric horn button is mounted in the centre of the steering wheel.

BODY CUM CHASSIS.

The vehicle as a whole is of unitary construction, there being no chassis frame in the accepted sense, the units being built into a basic platform which has a tunnel running from front to rear. It is a pity that the designers have found it necessary to use this tunnel, as it to a large extent offsets one of the advantages of a rear engine, i.e. the ability to use a flat floor. It is obvious however that the use of this construction has enabled a vehicle of a very light weight to be achieved, and in this direction the results are very satisfactory.

Regarding the bodywork itself, a separate report is being issued by the Body Experimental Department, and therefore reference will only be made to minor points insofar as they affect the driver or occupants whilst actually travelling or getting into the car.

From the driver's point of view, forward and downward vision is good, but the thick and steeply curved windscreen pillars cause a very serious blind spot.

Entry and exit to the front seats is satisfactory, two side doors being provided and hinged on the screen pillars.

The recessed pull out door handles are considered to be an attractive feature, although both doors require to be slammed very hard to make them close properly.

The seating positions both front and rear are good, though greater comfort would be desired by the normal British standard.

A cloth upholstery of a very poor quality is used, and this is already showing signs of wear.

There is a small luggage space behind the rear seat squab, and additional space is available in front of the driver in the position normally occupied by the engine.

The petrol tank is carried in this compartment, as also are the spare wheel and tools. Petrol fumes enter the car from the tank filler, the cap of which does not seal properly.

Access to the engine is gained through a hinged lid in the tail, which is provided with louvres for air cooling.

A considerable amount of thought has evidently been given to the question of heating the interior of the car, as a heating system is built in as part of the design. The hollow body side members are used as hot air ducts and ports are placed opposite to the front and rear passengers feet. There are also small ports at the bottom corners of the windscreen so that hot air may be blown on to the screen to act as a defroster. The heating system is controlled by means of a flap valve which bypasses the air either directly out to the rear of the vehicle when the heating system is not required, or through into the body of the car during cold weather. The flap valve is controlled by a small push/pull knob placed between the two front seats. This system, whilst being admirable in many ways, has the disadvantage of bringing with it smell and fumes from the engine.

WEIGHT.

In order to determine the effect on the front and rear

weight of the carrying of one or more passengers, the front, rear and total weights were checked in the unladen condition, and as each successive passenger was seated in various positions. Details of these weight figures are given below.

Effect of various loads on weight distribution.

Condition.	Front.	Rear.	Total.
Unladen	6-0-0 = 672 lbs.	7-3-21 = 889 lbs.	13-3-21 = 1561 lbs.
Driver only	6-3-21 = 777 "	8-2-7 = 959 "	15-2-0 = 1726 lbs.
Driver & front passenger	7-2-7 = 847 "	9-0-14 = 1022 "	16-2-21 = 1869 "
Driver, 1 front & 1 rear	7-3-14 = 882 "	10-0-21 = 1141 "	18-0-7 = 2023 "
Driver, 1 front & 2 rear.	8-0-14 = 910 "	11-0-21 = 1253 "	19-1-7 = 2163 "
Driver, 1 front & 3 rear.	8-2-0 = 952 "	12-1-0 = 1372 "	20-3-0 = 2324 "

COMPARATIVE TEST DATA.

	VOLKSWAGEN.	MINX.
1. WEIGHT. Complete vehicle unladen with petrol, oil and water, spare wheel and tools.		
Front	6 - 0 - 0 = 672 lbs.	9 - 0 - 21 = 1029 lbs.
Rear	7 - 3 - 21 = 889 "	9 - 1 - 14 = 1050 "
Total	13 - 3 - 21 = 1561 "	18 - 2 - 7 = 2079 "
2. ENGINE CAPACITY.		
C.C.s.	1161 c.c. = 70.8 cu.in.	1184.5 c.c. = 72.26 cu.in.
3. ACCELERATION FIGURES.		
2 up. Weight	16 - 2 - 13 = 1861 lbs.	20 - 3 - 0 = 2324 lbs.
Top gear 10-30	17.1 secs.	14.9 secs.
20-40	18.4 "	15.3 "
30-50	29.35 "	18.8 "
10-20	9.9 "	7.7 "
10-40	27.8 "	23.75 "
10-50	44.15 "	34.35 "
3rd gear 10-20	4.05 "	4.05 "
10-30	7.05 "	8.48 "
10-40	14.5 "	14.7 "
2nd gear 10-20	2.65 "	2.3 "
Rest to 50 using		
1st, 2nd, 3rd, Top	39.0 "	25.5 "
2nd, 3rd, Top	40.25 "	25.9 "
Rest to 40 using		
1st, 2nd, 3rd, Top	22.25 "	17.45 "
2nd, 3rd, Top	23.05 "	17.47 "
1st, 2nd, 3rd.	19.75 "	15.1 "
2nd, 3rd.	20.0 "	15.9 "

	VOLKSWAGEN.	MINX.
Rest to 30 using 1st, 2nd, 3rd, 2nd, 3rd.	13.5 secs. 13.6 "	9.7 secs. 9.6 "
Rest to 20 using 1st, 2nd. 2nd.	7.8 " 7.9 "	4.5 " 5.2 "
Rest to 10 using 1st. 2nd.	Speedometer needle erratic.	1.8 " 2.2 "
Standing 1/4 mils using: 1st, 2nd, 3rd, Top 2nd, 3rd, "	27.7 secs. 28.6 "	27.6 " 27.5 "
Mean max. speed timed 1/4 mile	56.3 mph.	61.5 mph.
4. HILL CLIMBING.		
<u>Frizz Hill. 2 up</u> <u>Weight.</u>	1861 lbs.	2324 lbs.
Start MPH.	40	40
Top MPH.	18	23.5
Time, secs.	41	39.5
Gear	Top	Top
<u>3 up</u> <u>Weight</u>	See 5 up + 160 lbs. weight.	2,475 lbs.
Start MPH.		40
Top MPH.	Not taken.	20.5
Time secs.		41.7
Gear		Top

	VOLKSWAGEN.	MINX.
4 up + 200 lbs. luggage, weight		2,823 lbs.
Start MPH.	)	40
Top MPH	) Not taken.	7
Time, secs.	)	63.3
Gear	)	3rd (change at 10 mph)
5 up + 160 lbs. luggage, weight.	2,471 lbs.	See 3 up weight.
Start MPH.	40	)
Top MPH	23	) Not taken.
Time, secs.	39.8	)
Gear	3rd.	)
Edge Hill.		
2 up weight	1861 lbs.	2324 lbs.
Start, MPH.	40	40
Change to 3rd @	30	29
Change to 2nd @	20	21
Top MPH.	25	28.5
Time, secs.	87.5	84.4
3 up weight	See 5 up + 160 lbs. weight	2475 lbs.
Start		40
Change to 3rd @	)	29
Change to 2nd @	)	21
Top mph.	) Not taken	27
Time secs.	)	88

	VOLKSWAGEN	MINX.
4 up + 200 lbs. luggage weight		2823 lbs.
Start MPH.		40
Change to 3rd @		28
Change to 2nd @	Not taken	21
Top MPH.		24
Time secs.		91.1
5 up + 160 lbs. luggage weight	2471 lbs.	See 3 up weight.
Start, MPH.	40	
Change to 3rd @	30	
Change to 2nd @	20	Not taken.
Top mph.	18	
Time, secs.	104.1	
5. **PETROL CONSUMPTION.**		
2 up weight at	1861 lbs.	2324 lbs.
Av. speed.	37.6 mph.	36 mph.
Consumption.	43 mpg.	34.6 mpg.
3 up weight		2475 lbs.
@ av. speed	Not taken.	36 mph.
Consumption		34.1 m.p.g.

	VOLKSWAGEN.	MINX.
4 up + 200 lbs. luggage, weight	)	2823 lbs.
At av. speed.	) Not taken.	36 mph.
Consumption.	)	31.2 m.p.g.
5 up + 160 lbs. luggage, weight	2471 lbs.	)
@ av. speed.	36.3 mph.	) Not taken.
Consumption.	40 mpg.	)

NOTE.

1. Volkswagen 5 up + 160 lbs. comparable in weight with Minx 3 up.

2. Atmos. temp. varied from 40°F to 62°F during tests.

3. Barometer varied from 29.6" to 30.03" during test.

6. BRAKE PERFORMANCE

Normal fully laden weight	5 up = 2311 lbs.	4 up + 150 lbs. = 2773 lbs.
Pedal pressure lbs.	Deceleration %.	Deceleration %.
20	4.65	20
40	18.65	42
60	29.5	64
80	42	82
100	48	88
120	54.3	
140	60.5	
160	66.75	
180	68.3 front wheels just locked.	

	VOLKSWAGEN	MINX.
7. Turning circle		
R.H.	32' 9"	34' 6"
L.H.	36' 0"	34' 6"
8. Wheelbase	7' 11½"	7' 8"
9. Track Front	4' 3"	3' 11½"
Rear	4' 0"	4' 0½"
10. Tyres	5.00 - 16	5.00 - 16
11. Overall height unladen	5' 0"	5' 2½"
12. Overall length	13' 1"	12' 11¼"
13. Overall width	5' 1"	5' ½"

PART III.

BODY ENGINEERING

of the

VOLKSWAGEN SALOON, TYPE 11.

Report by:-

HUMBER LTD.

INDEX TO ILLUSTRATIONS.

ILLUSTRATION NO.	TITLE.
1.	Dimension chart.
2.	Breakdown of body panels - front end.
3.	Breakdown of body panels - body and rear end.
4.	Key drawing to sections.
5.	Sections front wing and rear roof panel.
6.	Section floor and body sill.
7.	Section "A" and "BC" posts and door.
8.	Sections - sealing engine compartment cover - mud draft front wing and sealing on bonnet.
9.	Section luggage compartment floor and sketch trim pad clip.
10.	Sections - roof panel.
11.	Door lock mechanism.
12.	Window lift mechanism.
13.	Bonnet and engine compartment lock.
14.	Bonnet support mechanism.
15.	Engine Compartment Hinge.
16.	Rear seal construction.
17.	Heating system.
18.	Front three quarter view of vehicle.
19.	Rear three quarter view of vehicle.
20.	Side view of vehicle.
21.	Front view of vehicle.
22.	Close up of engine layout.
23.	Rear engine compartment.
24.	Facia and dash arrangement.
25.	Front end arrangement.
26.	Arrangement of controls.
27.	Front floor and axle arrangement.

----ooOoo----

VOLKSWAGEN SALOON.

General notes on vehicle with wartime finish, inspected by the Body Engineering Department.

General Data.

The vehicle inspected was a 2-door four light fixed roof Saloon with accommodation for five passengers, two on front bucket seats and three on the rear seat.

Wheelbase	94	
Front Track	51.0	
Rear Track	48.0	
Overall length	159.50	
Overall height	59.5 (unladen)	
Overall width	60.50	
Front Axle weight	6. 0. 0	
Rear Axle weight	7.3.21	
Total Weight	13. 3. 21	
(Unladen but including petrol and oil)		

6 volt lighting equipment with battery provided with sheet metal cover fitted under the forward edge of the rear seat. Recessed headlamps, incorporating head and dipped arrangement, together with sidelamp bulbs, Two tail lamps are provided, one on each rear wing, and a centre lamp on the engine compartment lid for the number plate. A remote gear change control is fitted centrally to the backbone tunnel, with a horizontal type handbrake lever placed behind it, between the front bucket seats.

(For Dimension Chart see illustration No. 1.)

BODY DETAILS.

BODY SHELL.

Fully pressed steel panel construction of advanced design is employed, and is based on the semi-unitary principle, i.e., the body shell is used to reinforce the chassis or underframe assembly. The main strength of the underframe structure is obtained by the use of a completely closed ⌒ channel, which houses the control rods and cables, and forms the backbone of the platform to which, at front and rear, suitable outrigger members are attached for carrying the suspension, and the rear mounted horizontally opposed power unit. The main panel pressings are large, and the tooling has been designed to produce large complete panels, often of a complex type, i.e., the front and rear outside panels also include the wheelarch pressings, (see sketch showing the panel split up) with a view to eliminating fabrication, particularly spot and flash welding during the assembly of the body. Certain main body joints have been made by flanging, thus further eliminating spot welding. This applies to the fixing of the roof panel to the side panels, where the water gutter is flanged over its edge, also to the front outside panels where they are attached to the inside front reinforcement pressing, and to the rear side panels where attached to the inside reinforcement panels. (See sections No. 11, No. 4, No. 5). The panel split up is exceedingly interesting, and is briefly as follows for the main panels:-

(a) Front side pressings, these include the wheelarches and a flanging for picking up the inside front reinforcement pressing.

(b) Rear side pressings, these include the quarter lights, the rear wheelarches which are suitably flanged at their bottom edges to act as a reinforcement, the forward face of the door shut pillar and a flanging for picking up with the inner reinforcement panel.

(c) The roof panel, which extends from the back edge of the front compartment lid to the rear engine compartment lid and includes the winscreen aperture, twin backlights, and cooling air grille.

(d) The front inner reinforcement which includes the screen header inner reinforcement, the facia panel and also the faces of the front hinge pillar.

(e) The side panel reinforcements which extend from the front inner reinforcement to the extreme back of the body behind the rear seat squab, and which also form the equivalents of cantrails over the doors.

(f) Front and engine compartment lids.

(g) The front and rear centre joining panels, the front compartment floor, and the sill panels.

(For breakdown of body panels, see illustration No. 2 and 3)

It was noted that the facia panel pressing was so designed that it was suitable for L.H. and R.H. drive, and that the demister nozzles were built into it at the corners of the windscreen; the air ducts for these and heating being built into the box sections, forming the body shell structure. The inner reinforcements, though painted in wartime finish are not unsightly, and could with suitable finish be made to look quite attractive.

(For sections of body shell, see key illustration No. 4. and
illustration No. 5 - 10)

THE DOORS.

These are hinged at their forward edges on two leaf hinges with
plain pins and one leaf formed into a cover for weather protection.
The doors are made up of two main pressings spot welded together,
(see illustration No. 7) and the gap between the two pressings for
accomodating the glass channel is bridged locally with brackets.
The flanges condition for spot welding the outer and inner panels
together also forms a retainer into which a rubber seal is
effectively secured, and a deep check is provided in the hinge and
shut pillars to accommodate this. The doors are provided with
locks fitted with a rotary type of striker to give easy closing,
"Pullout" type handles are fitted, (the RH outside handle incorporates a lock cylinder) see illustration No. 11. Remote controls
bring the inside handles conveniently forward to suit the front
passengers. The doors are provided with two large composition
'dovetail' type buffers which fit into recesses provided in the
check of the shut pillar; no metal dovetail of any sort is provided.
The design of the window winder as can be seen from illustration
No. 12 is of particular interest, this fitting is of the parallel
arm type and is exceedingly low geared with the usual type of
clutch mechanism incorporated. The attachment bracket is made
from strip pierced and formed for either R.H. or L.H. assemblies,
the whole plungings for taking the arms to suit either conditions
being provided. These plungings are approximately $\frac{1}{2}$" in diameter
and when "peened" over on the arms afford a much larger bearing
condition than is usual. Another feature of this plunging is that
it incorporates a secondary plunging at its base which forms a
washer surface for the arms. The blanking of the toothed quadrant
has also been most carefully considered and inspection shows that
it has been nested out of strip to the extent of the blanking marks
from one showing in the other. Inside door and winder handles are
retained on spline fixings by grooved pins.

The glass channels, which as can be seen from the section drawing
(illustration No. 7) are of particularly heavy section and comprise
a steel beaded channel with rubber insert which is covered with a
black imitation moquette material. Door lights are of safety glass
Door insulation consists of paper felt of the cellular type and the
door trim pad is of hardboard padded and covered with leather cloth,
clips (see illustration No. 9), being provided for fixing this to
the inner door panel. Metal check straps are fitted and function
effectively, allowing the doors to open to approx. 90°.

THE QUARTER.

The quarterlights are of "Perspex" retained in a well designed rubber section which in turn is secured by the spot welding flange formed between the inner and outer panels (see section No. 4). The space between the reinforcement is provided with a trim pad covered with leathercloth fixed by means of clips to the side reinforcements.

THE ROOF.

This consists of a single pressing and extends from the edge of the front compartment lid to the engine access lid and includes the screen aperture, two back-lights, the cooling air grille for the engine fan, and at the front includes a well designed water gutter for the luggage compartment lid.

No insulation or anti-drum is fitted to the roof panel and an original method of fitting the roof lining is employed. The cloth is stitched to a number of spring steel rods which include millboard retaining strips, the whole thus being fitted into the recess between the roof panel and the inner side reinforcement of the body, (see section No. 12)

WINDSCREEN AND FACIA.

The Windscreen is of safety glass fixed in a rubber seal of similar design to that used for the quarterlights, and is provided with double non-parking wipers at the bottom edge. As previously mentioned defroster slots are provided at the two lower corners, and the ducts for these are incorporated in the 'A' post pressings. An adjustable rear vision mirror is fitted centrally at the top of the screen.

The Facia Panel is provided with an instrument panel of extremely simple design consisting of speedometer, headlamp switch, screen wiper switch and rear interior light switch combined, and ignition and oil tell-tale indicators. A symetrically opposite recess is provided for the instrument panel on R.H. drive, and is filled when not required by a pressed panel with the maker's badge embossed on to it. Two reasonable sized cubby boxes are provided one at each side.

A point of interest here is the ample leg room provided by this arrangement for the front passengers.

FRONT SPARE WHEEL AND REAR ENGINE COMPARTMENTS.

The design of the lids for both these compartments is of considerable interest, both being comparatively large but light pressings with suitable inner reinforcements which incorporate flanged lightening holes giving additional strength to them. Adequate rubber seals are provided, these being fitted in metal retainers (see illustration No. 8)

The front lid is provided with internal concealed type swan necked hinges made up of two pressings spot welded together.

It has a catch of simple yet effective design operated by an outside locking handle (see illustration No. 13) and a self acting support is provided for keeping the lid open. The action of this support is such that when the lid is opened it automatically engages and the operation of the lifting movement causes it to disengage thus allowing the closing of the lid (see illustration No. 14). The front compartment contains a well for carrying the spare wheel, the petrol tank, the tool kit and space for a small amount of luggage. Access is also easily attainable to the back of the instrument panel, and to the screenwiper motor and linkage.

The rear engine compartment lid is of similar construction to the front, including a catch and locking handle, is hinged on concealed type hinges and spring balanced by means of a round torsion spring which moves over centre, (see illustration No. 15).

THE FLOOR.

This is actually part of the chassis or underframe assembly, and consists of a flat reinforced platform of 20 swg. with stiffening swages, and a central closed section which forms a backbone for this, and to which at front and rear outriggers are provided for the attachment of the suspension and power unit.
The body is mounted to this platform by means of bolts at its outside edges, and a rubber insulation is provided between the body and the platform. The bottom side member is of box section and this is utilised in the heating system and as a duct for the passage of warm air to the passenger compartment. The body structure, particularly because of the sill construction considerably reinforces this platform and together an exceptionally rigid and semi-unitary vehicle is formed.
The floor is covered with a rubber mat and is not provided with any form of underlay.
The side body sills and front of body forming the front passengers foot space are covered with haircord carpet, care being taken to retain the carpet by means of a metal tread strip in the door openings.
(See illustration No. 6 for section through floor.)

SEATS.

Front Seats.

These are of the flat backed bucket type with tubular frames, and hinged back-rest to give easy access to rear seat. Tension springs are used for both cushion and squab, the cushion also being provided with a spring and hair topped pad; while only hair padding is provided for the backrest. Both seats are provided with adjustment which takes the form of a simple wing nut and clamping device.

Rear Seat.

This gives ample and comfortable room for three people. The seat cushion and squab of the springweave type mounted to a tubular frame with hair padding to the top and covered like the front seats with wool cloth. The cushion is removable and rests on simple ledges provided at the sides of the body and on the centre tunnel. The squab is hinged at its lower edges and rests against rubber covered pegs fitted to the body side reinforcements. When hinged forward access can be easily attained to the space **behind** it **for carrying luggage** (see illustration No. 16)

THE REAR LUGGAGE COMPARTMENT.

This space behind the rear seat squab is provided with special insulation to the floor and the bulkhead between itself and the engine compartment. This insulation consists of one layer of paper felt, one layer of sizalcraft paper, one layer of jute felt, and a top layer of black, smooth finished, embossed millboard with wood protecting battens. (See illustration No. 9)

BACKLIGHTS.

Two backlights are provided, these being included in the main roof pressing. These are of safety glass and fixed into the body by means of a similar rubber section to the screen and quarterlights.

WINGS.

These are one piece pressings bolted to the side panels of the body shell. They are of clean design, the front incorporating the recessed head-lamp pressings, and the rear pressings for carrying the tail lamps, and on both wings the side cutting line is worked up into a 'J' section which acts as an effective mud trap (See section No. 7)

SIDE SILLS.

Side sills are provided between the wings and are attached to the side of the body shell. These act to keep down mud splash, and are rigidly braced with suitable stiffeners.

FRONT AND REAR BUMPERS.

These are steel 16 swg. pressings with spring steel backbars and overriders. Like the rest of the vehicle they are in painted wartime finish.

---ooOoo---

HEATER AND DEMISTER SYSTEM.

The system is of considerable interest; it utilises the hot air from the engine and the pressure built up by the air cooling fan for maintaining its circulation. The ducts for this system, as already mentioned, are built into the bodyshell structure, and the control of it is by means of a "butterfly" valve operated by means of a cable with push-pull knob situated on the centre tunnel between the front seats. Four heating vents are provided at either side of the body alongside the front and rear passenger's feet. As previously mentioned the de-froster slots are incorporated in the facia panel at the windscreen corners. (See illustration No.17).

TRAFFICATORS.

Provision is made for these in the body side panels and a raised surface is provided on the facia pressing for the fitting of the switch, together with red and green tell-tale lamps at the R.H. of the instrument panel. No trafficators or switch, however, are actually fitted on the vehicle inspected.

GENERAL.

The car is painted entirely in Army Green, and this is an exceedingly poor finish. No adequate measures seem to have been taken either for cleaning and de-greasing the pressings before priming, or any steps taken for rustproofing, with the result that the paintwork is peeling away from the metal.

It is our considered opinion that from the Body Engineering point of view the design of this vehicle is exceptionally good, and shows a great advance on previous constructional methods, but workmanship and general finish of the vehicle leave much to be desired and could be improved.

LIST OF BODY AND PANEL GAUGES.

Location.	Gauge.
Door Panels	21 swg.
Roof Panel	21 swg.
Front Lid	21 swg.
Engine Compartment Lid	22 swg.
Side Panels (Inner & Outer)	20 swg.
Front Reinforcement	22 swg.
Wings	19 swg.
Floor	18 swg.
Floor Tunnel	14 swg.
Front and Rear Lid Reinforcements	22 swg.
Floor Reinforcements	18 swg.

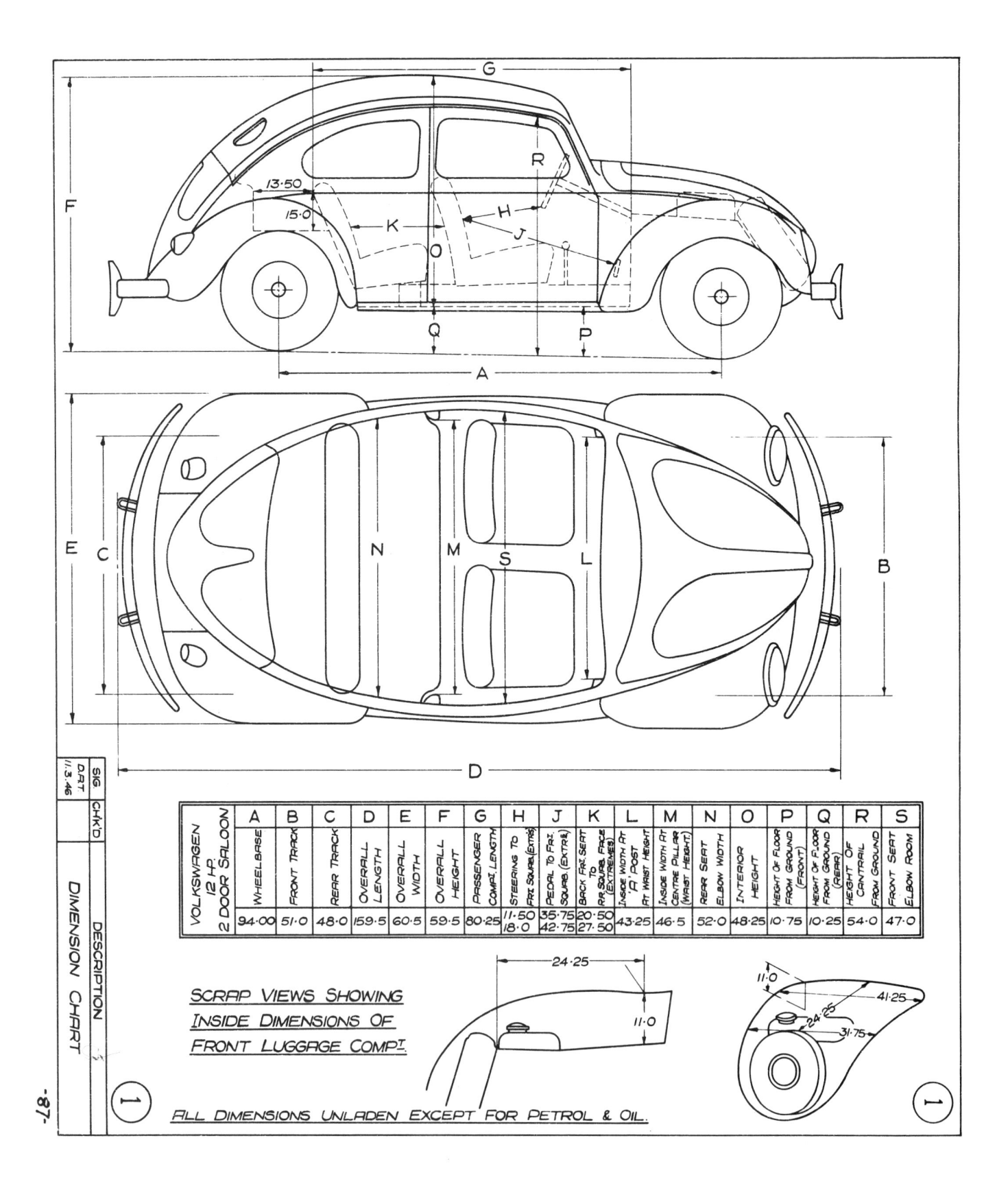

	A	B	C	D	E	F	G	H	J	K	L	M	N	O	P	Q	R	S
VOLKSWAGEN 12 H.P. 2 DOOR SALOON	WHEELBASE	FRONT TRACK	REAR TRACK	OVERALL LENGTH	OVERALL WIDTH	OVERALL HEIGHT	PASSENGER COMPT. LENGTH	STEERING TO FRT. SQUAB.(EXTRS)	PEDAL TO FRT. SQUAB. (EXTRS)	BACK FRT. SEAT TO RR. SQUAB. FACE (EXTREMES)	INSIDE WIDTH AT 'A' POST	INSIDE WIDTH AT CENTRE PILLAR (WAIST HEIGHT)	REAR SEAT ELBOW WIDTH	INTERIOR HEIGHT	HEIGHT OF FLOOR FROM GROUND (FRONT)	HEIGHT OF FLOOR FROM GROUND (REAR)	HEIGHT OF CENTRAIL FROM GROUND	FRONT SEAT ELBOW ROOM
	94.00	51.0	48.0	159.5	60.5	59.5	80.25	11.50 18.0	35.75 42.75	20.50 27.50	43.25	46.5	52.0	48.25	10.75	10.25	54.0	47.0

SCRAP VIEWS SHOWING INSIDE DIMENSIONS OF FRONT LUGGAGE COMP^T.

ALL DIMENSIONS UNLADEN EXCEPT FOR PETROL & OIL.

DIMENSION CHART

-87-

INTERIOR PANEL ROOF, SIDES ETC. NOT SHOWN.

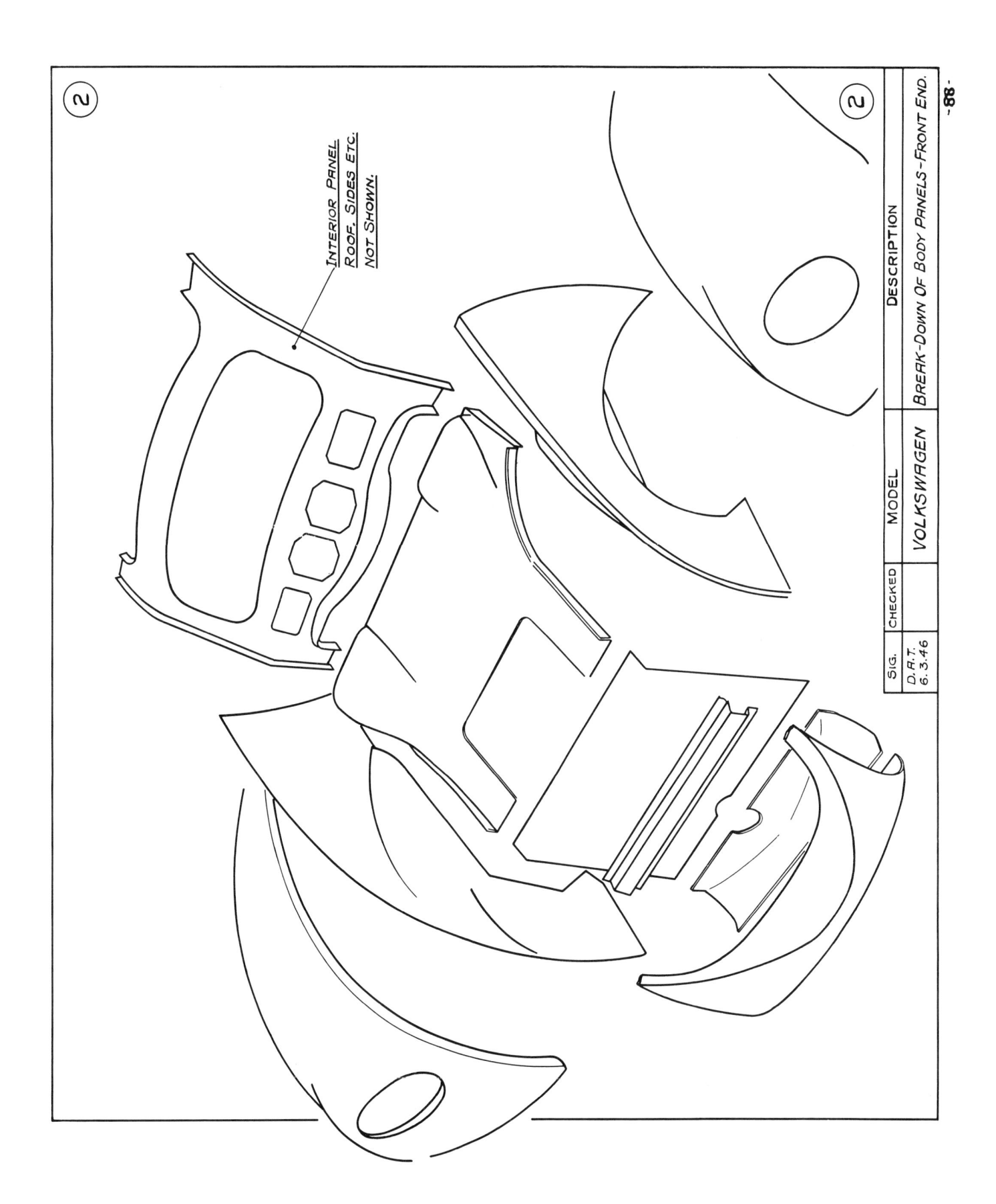

		MODEL	DESCRIPTION
SIG.	CHECKED	VOLKSWAGEN	BREAK-DOWN OF BODY PANELS - FRONT END.
D.R.T. 6.3.46			

OUTER SIDE PANEL.

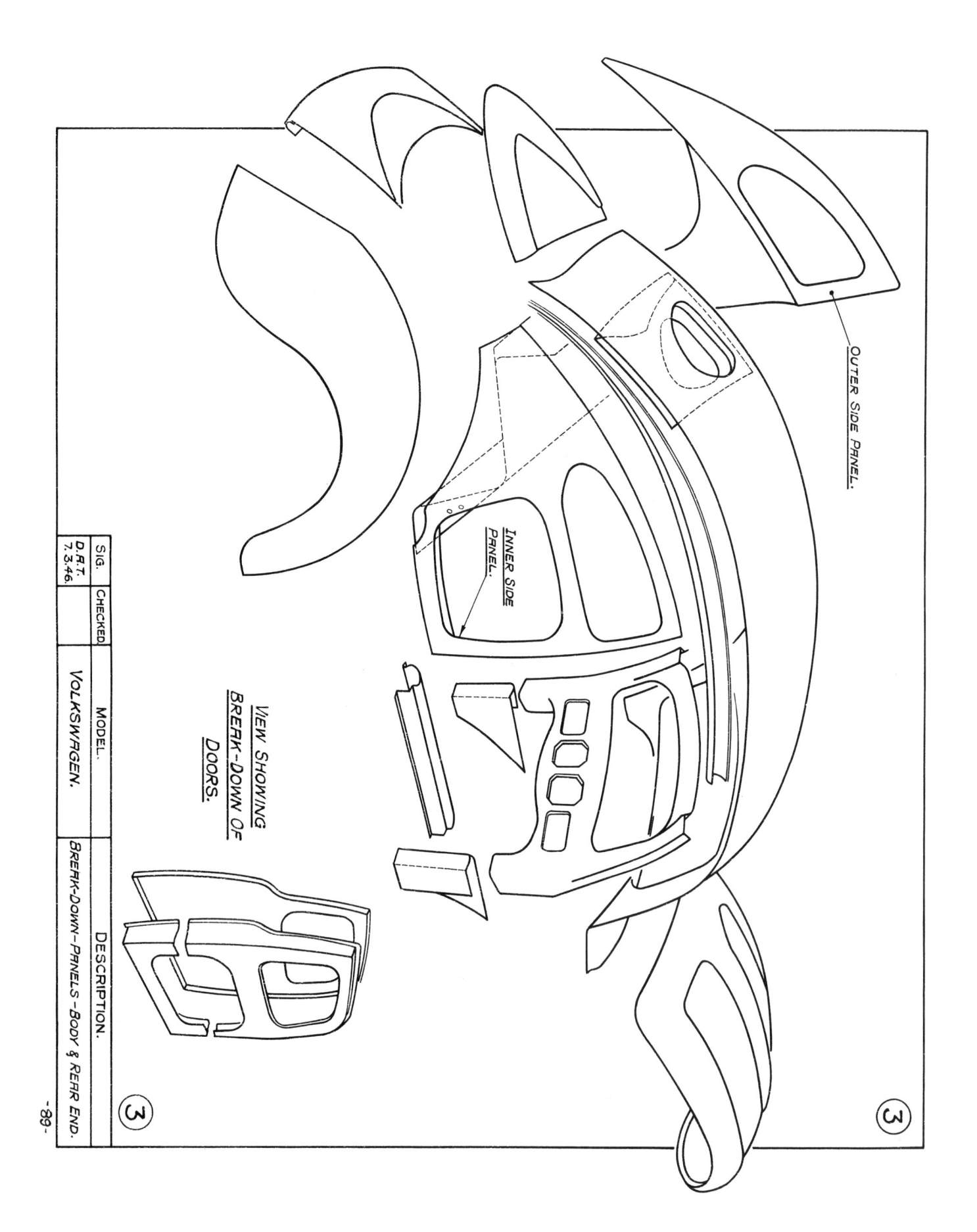

INNER SIDE PANEL.

VIEW SHOWING BREAK-DOWN OF DOORS.

SIG.	CHECKED	MODEL.	DESCRIPTION.
D.R.T. 7.3.46.		VOLKSWAGEN.	BREAK-DOWN-PANELS-BODY & REAR END.

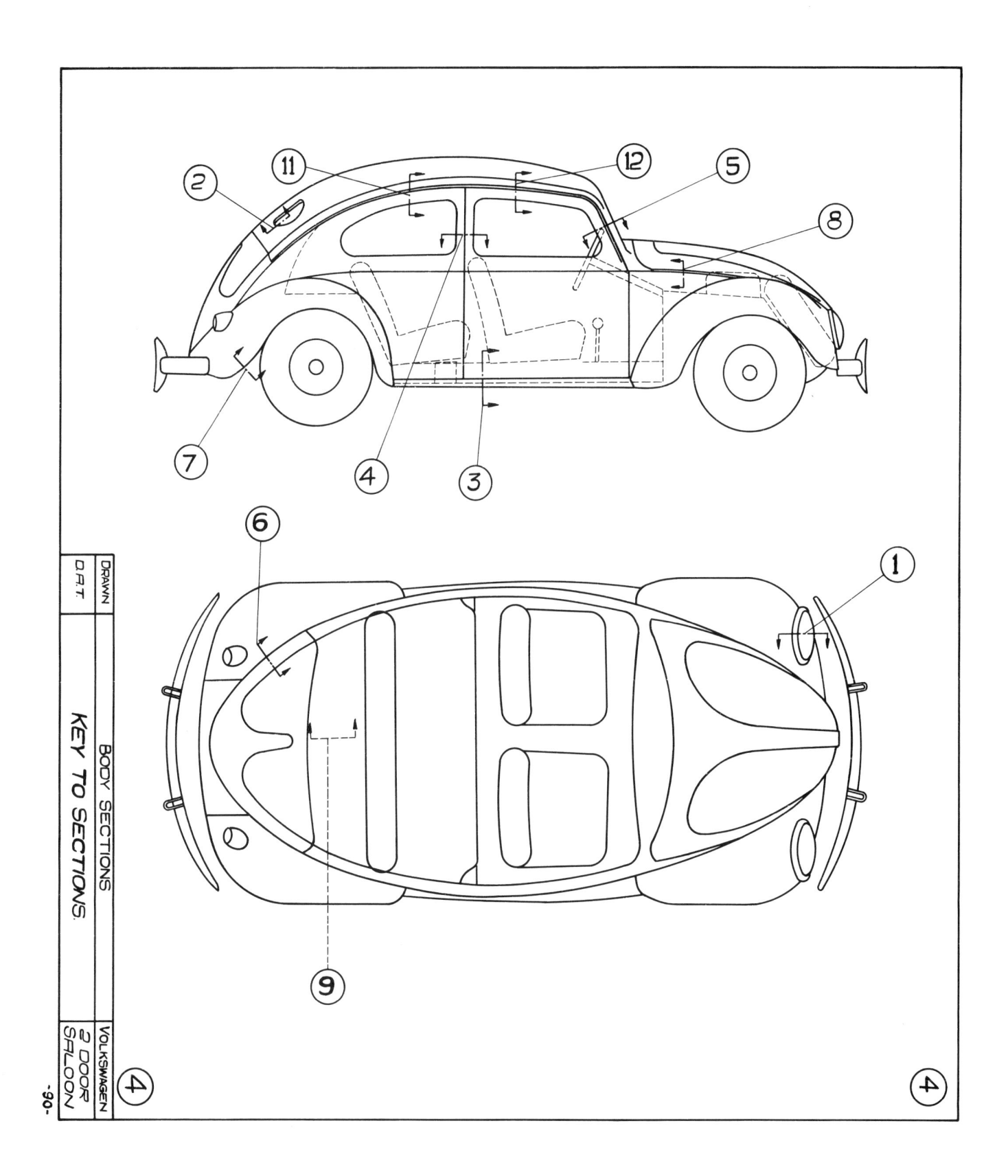

BODY SECTIONS
KEY TO SECTIONS

VOLKSWAGEN
2 DOOR SALOON

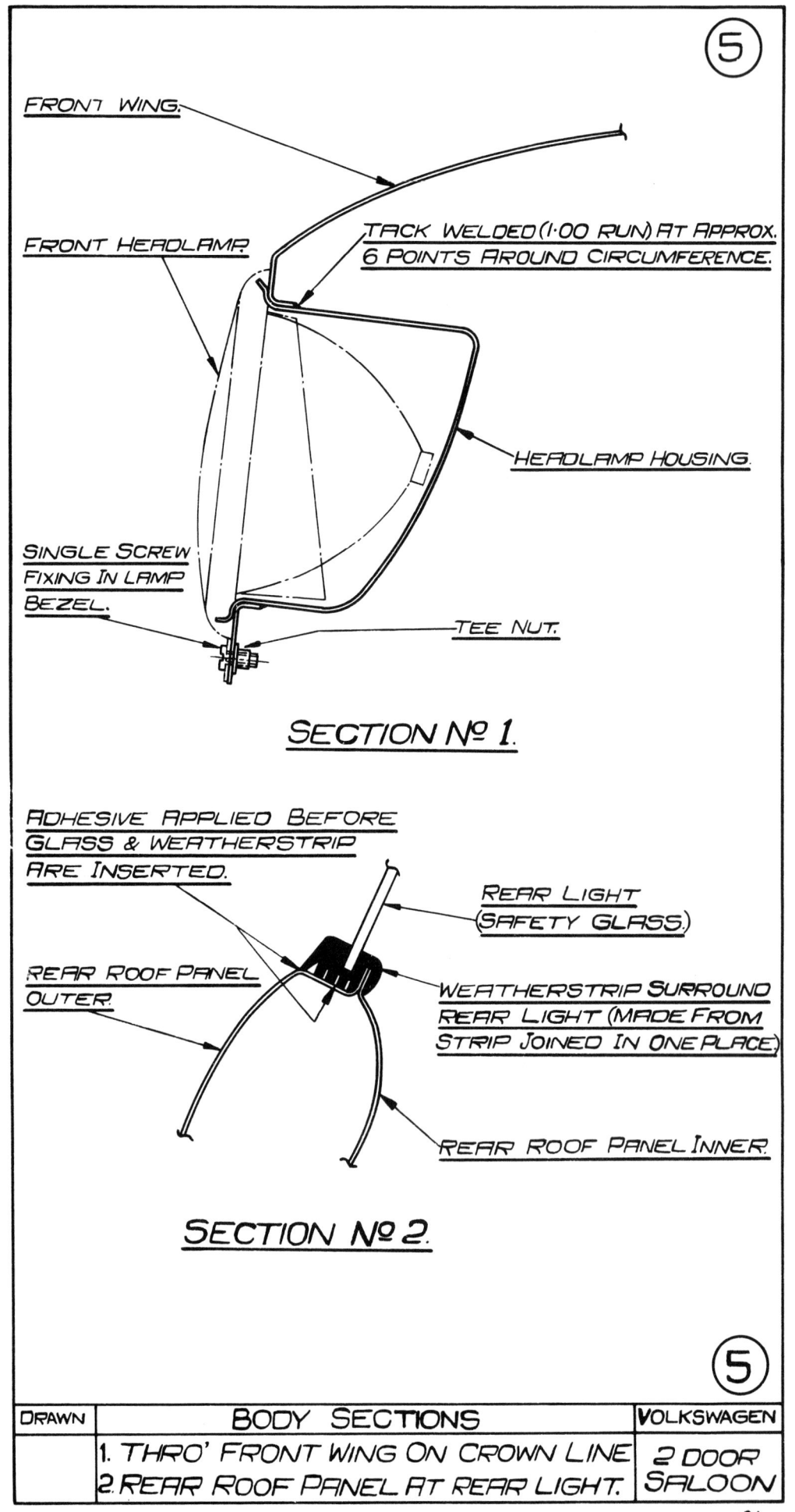

SECTION Nº 1.

- FRONT WING.
- FRONT HEADLAMP.
- TACK WELDED (1·00 RUN) AT APPROX. 6 POINTS AROUND CIRCUMFERENCE.
- HEADLAMP HOUSING.
- SINGLE SCREW FIXING IN LAMP BEZEL.
- TEE NUT.

SECTION Nº 2.

- ADHESIVE APPLIED BEFORE GLASS & WEATHERSTRIP ARE INSERTED.
- REAR LIGHT (SAFETY GLASS.)
- REAR ROOF PANEL OUTER.
- WEATHERSTRIP SURROUND REAR LIGHT (MADE FROM STRIP JOINED IN ONE PLACE.)
- REAR ROOF PANEL INNER.

DRAWN	BODY SECTIONS	VOLKSWAGEN
	1. THRO' FRONT WING ON CROWN LINE 2. REAR ROOF PANEL AT REAR LIGHT.	2 DOOR SALOON

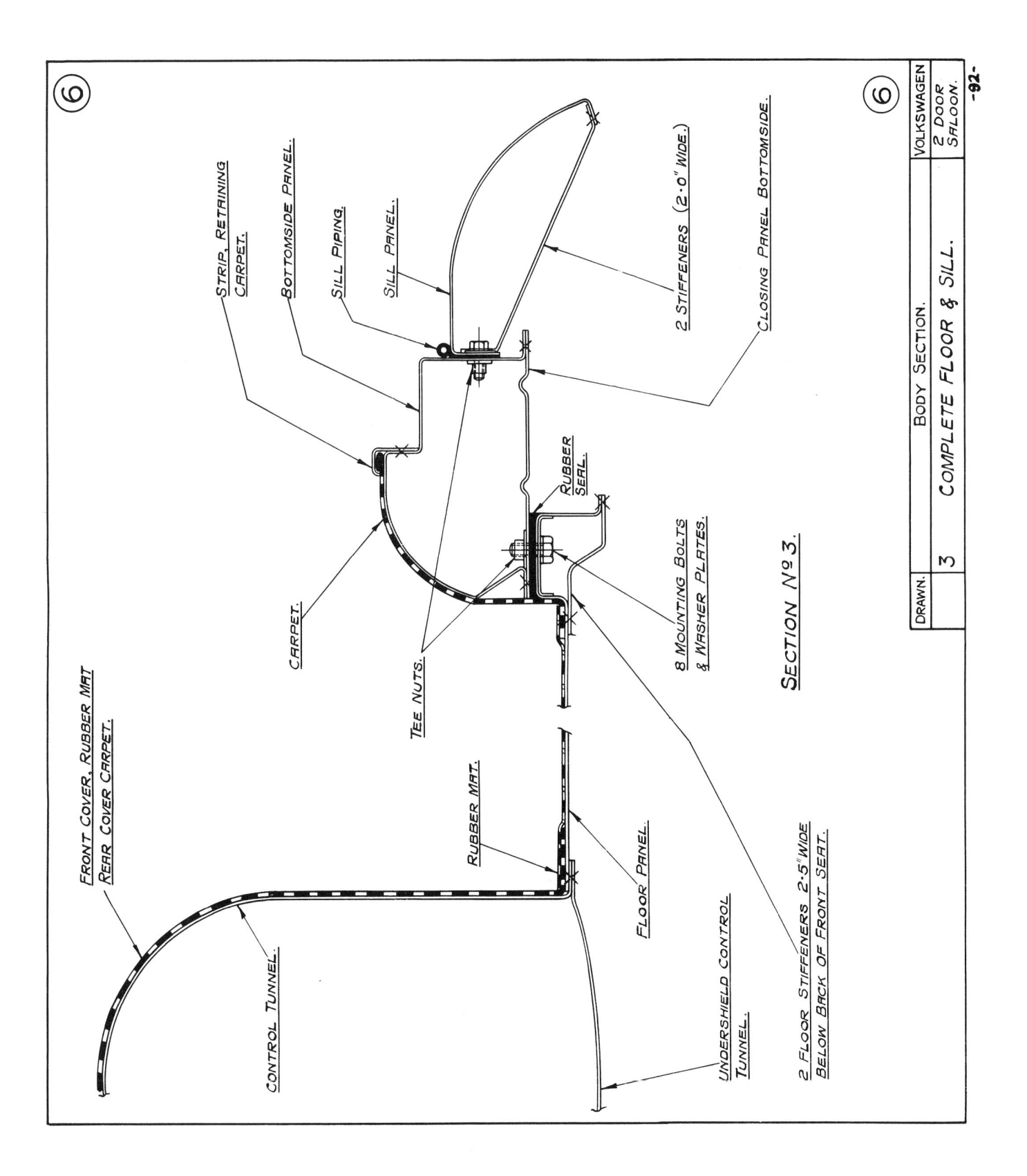

SECTION No. 3

Labels (clockwise from top):
- STRIP, RETAINING CARPET.
- BOTTOMSIDE PANEL.
- SILL PIPING.
- SILL PANEL.
- 2 STIFFENERS (2·0" WIDE.)
- CLOSING PANEL BOTTOMSIDE.
- RUBBER SEAL.
- 8 MOUNTING BOLTS & WASHER PLATES.
- TEE NUTS.
- CARPET.
- FRONT COVER, RUBBER MAT REAR COVER CARPET.
- CONTROL TUNNEL.
- UNDERSHIELD CONTROL TUNNEL.
- RUBBER MAT.
- FLOOR PANEL.
- 2 FLOOR STIFFENERS 2·5" WIDE BELOW BACK OF FRONT SEAT.

DRAWN.	BODY SECTION.	VOLKSWAGEN
3	COMPLETE FLOOR & SILL.	2 DOOR SALOON.

-92-

SECTION N° 4.

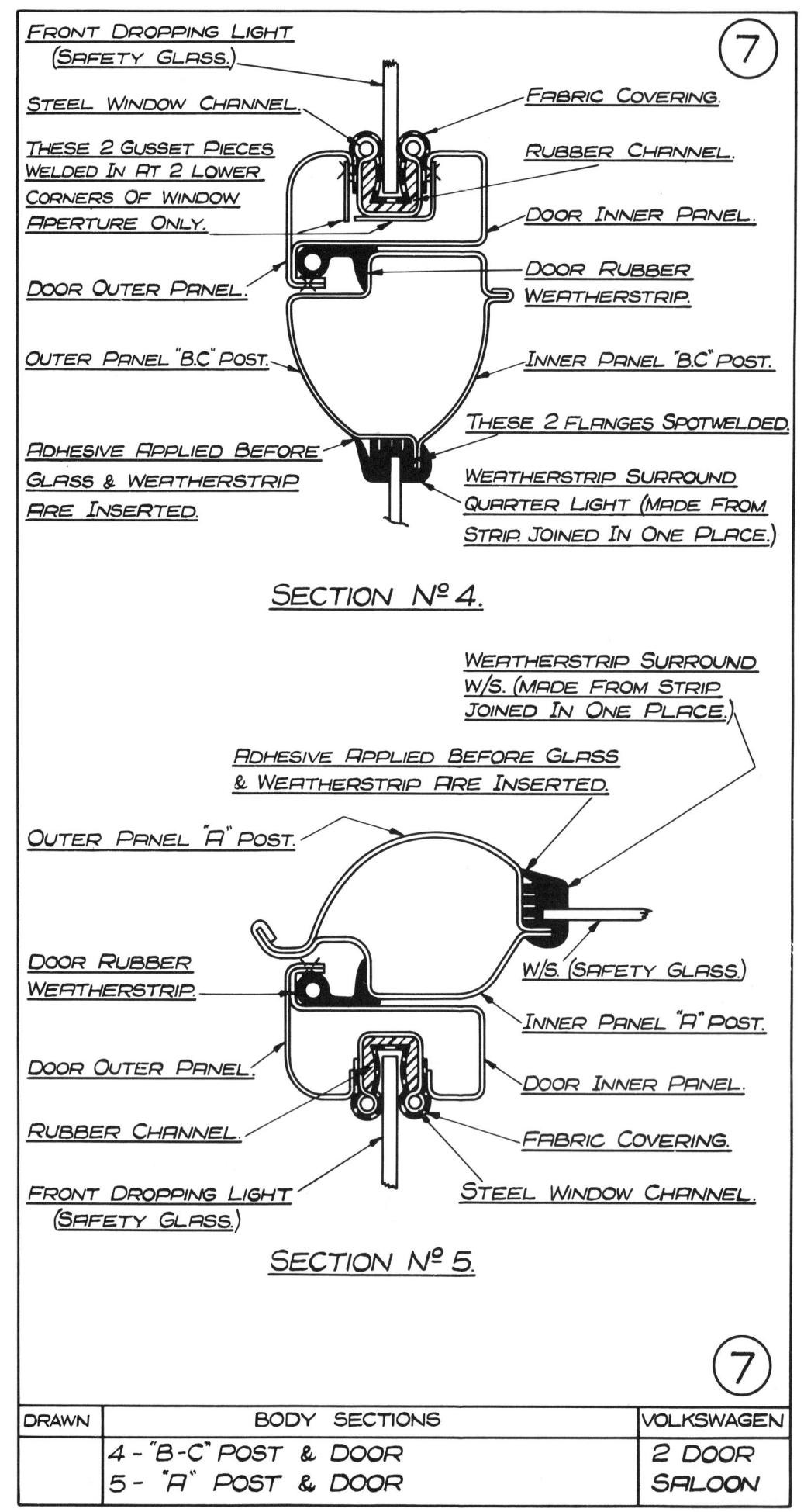

- FRONT DROPPING LIGHT (SAFETY GLASS.)
- STEEL WINDOW CHANNEL.
- THESE 2 GUSSET PIECES WELDED IN AT 2 LOWER CORNERS OF WINDOW APERTURE ONLY.
- DOOR OUTER PANEL.
- OUTER PANEL "B.C" POST.
- ADHESIVE APPLIED BEFORE GLASS & WEATHERSTRIP ARE INSERTED.
- FABRIC COVERING.
- RUBBER CHANNEL.
- DOOR INNER PANEL.
- DOOR RUBBER WEATHERSTRIP.
- INNER PANEL "B.C" POST.
- THESE 2 FLANGES SPOTWELDED.
- WEATHERSTRIP SURROUND QUARTER LIGHT (MADE FROM STRIP. JOINED IN ONE PLACE.)

SECTION N° 5.

- WEATHERSTRIP SURROUND W/S. (MADE FROM STRIP JOINED IN ONE PLACE.)
- ADHESIVE APPLIED BEFORE GLASS & WEATHERSTRIP ARE INSERTED.
- OUTER PANEL "A" POST.
- DOOR RUBBER WEATHERSTRIP.
- DOOR OUTER PANEL.
- RUBBER CHANNEL.
- FRONT DROPPING LIGHT (SAFETY GLASS.)
- W/S. (SAFETY GLASS.)
- INNER PANEL "A" POST.
- DOOR INNER PANEL.
- FABRIC COVERING.
- STEEL WINDOW CHANNEL.

DRAWN	BODY SECTIONS	VOLKSWAGEN
	4 - "B-C" POST & DOOR 5 - "A" POST & DOOR	2 DOOR SALOON

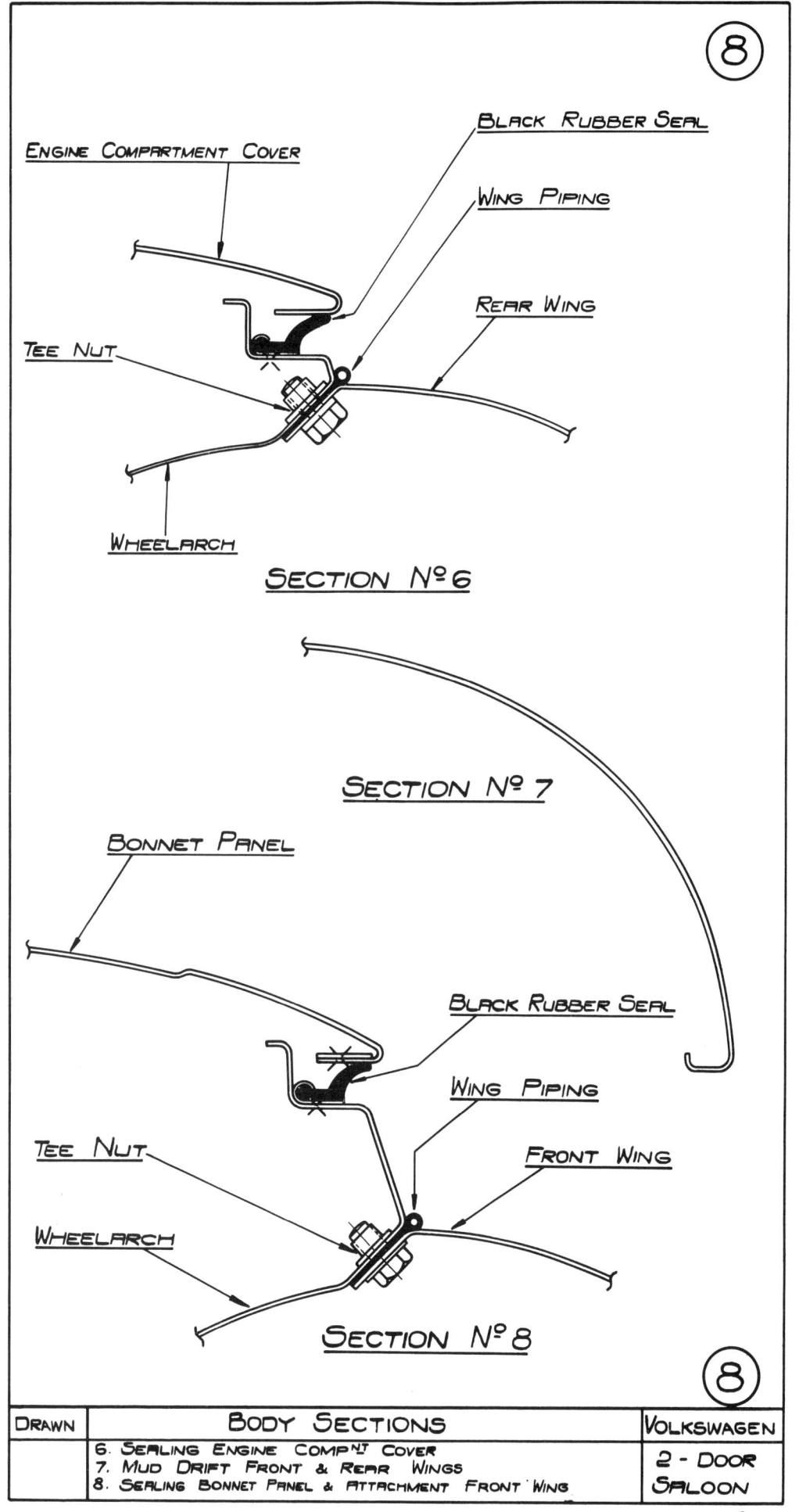

SECTION № 6

- ENGINE COMPARTMENT COVER
- BLACK RUBBER SEAL
- WING PIPING
- REAR WING
- TEE NUT
- WHEELARCH

SECTION № 7

SECTION № 8

- BONNET PANEL
- BLACK RUBBER SEAL
- WING PIPING
- TEE NUT
- FRONT WING
- WHEELARCH

DRAWN	BODY SECTIONS	VOLKSWAGEN
	6. SEALING ENGINE COMPT COVER 7. MUD DRIFT FRONT & REAR WINGS 8. SEALING BONNET PANEL & ATTACHMENT FRONT WING	2 - DOOR SALOON

⑨

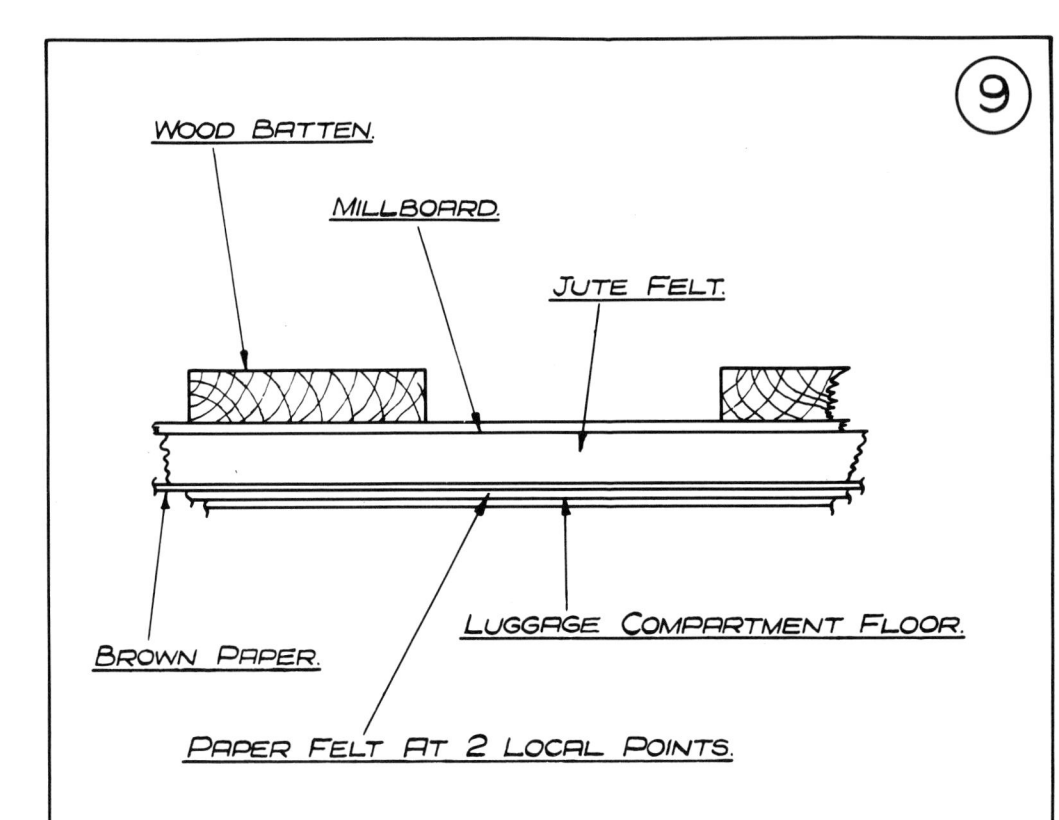

WOOD BATTEN.
MILLBOARD.
JUTE FELT.
BROWN PAPER.
LUGGAGE COMPARTMENT FLOOR.
PAPER FELT AT 2 LOCAL POINTS.

SECTION Nº 9.

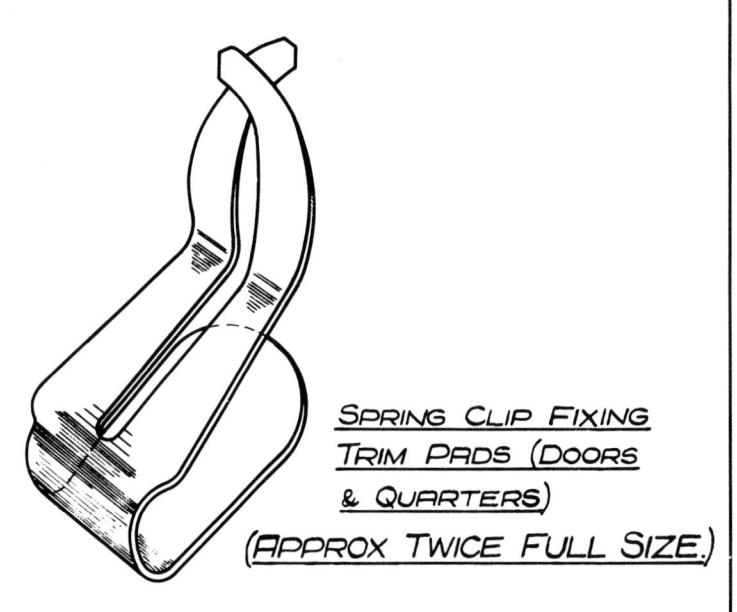

SPRING CLIP FIXING TRIM PADS (DOORS & QUARTERS)
(APPROX TWICE FULL SIZE.)

SKETCH Nº 10.

⑨

DRAWN	BODY SECTIONS	VOLKSWAGEN
	9. LUGGAGE COMPARTMENT FLOOR (INSULATION)	2 DOOR
	10. SKETCH TRIM PAD CLIP	SALOON

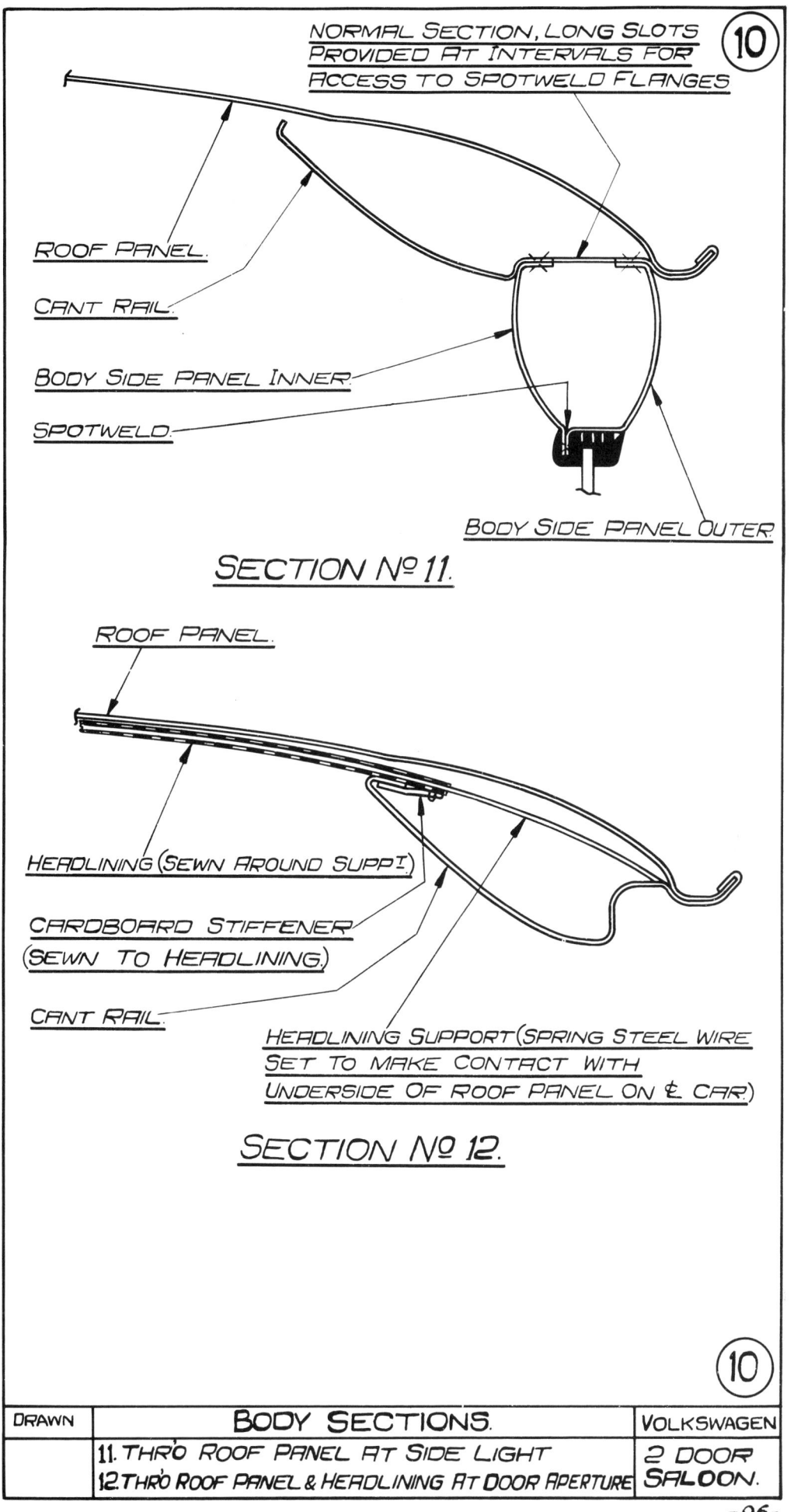

SECTION No. 11.

Normal section, long slots provided at intervals for access to spotweld flanges

- Roof Panel
- Cant Rail
- Body Side Panel Inner
- Spotweld
- Body Side Panel Outer

SECTION No. 12.

- Roof Panel
- Headlining (Sewn Around Suppt.)
- Cardboard Stiffener (Sewn to Headlining)
- Cant Rail
- Headlining Support (Spring Steel Wire set to make contact with underside of Roof Panel on ℄ car)

DRAWN	BODY SECTIONS.	VOLKSWAGEN
	11. THRO' ROOF PANEL AT SIDE LIGHT	2 DOOR SALOON.
	12. THRO' ROOF PANEL & HEADLINING AT DOOR APERTURE	

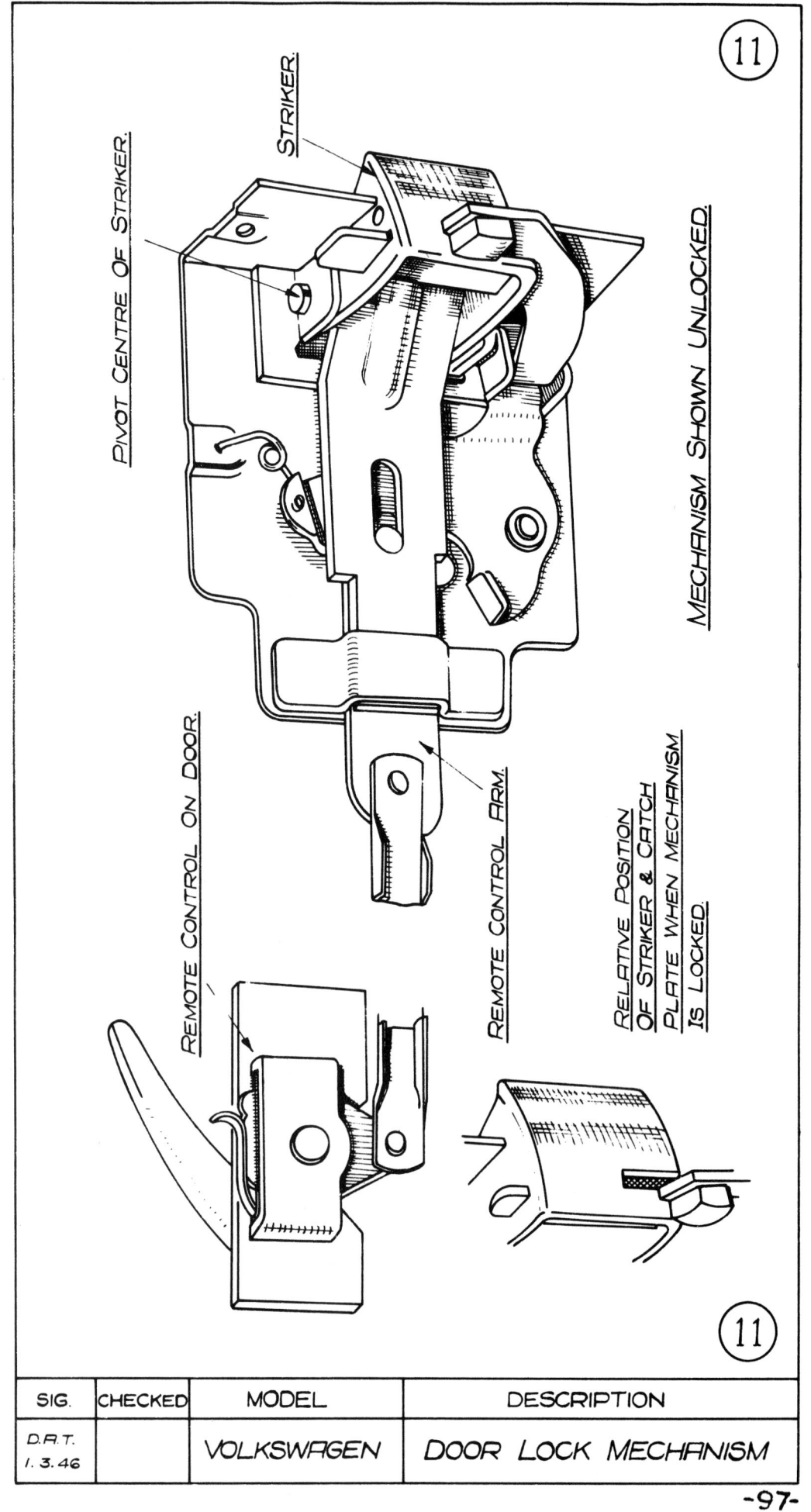

SIG.	CHECKED	MODEL	DESCRIPTION
D.A.T. 1.3.46		VOLKSWAGEN	DOOR LOCK MECHANISM

Window Lift Mechanism

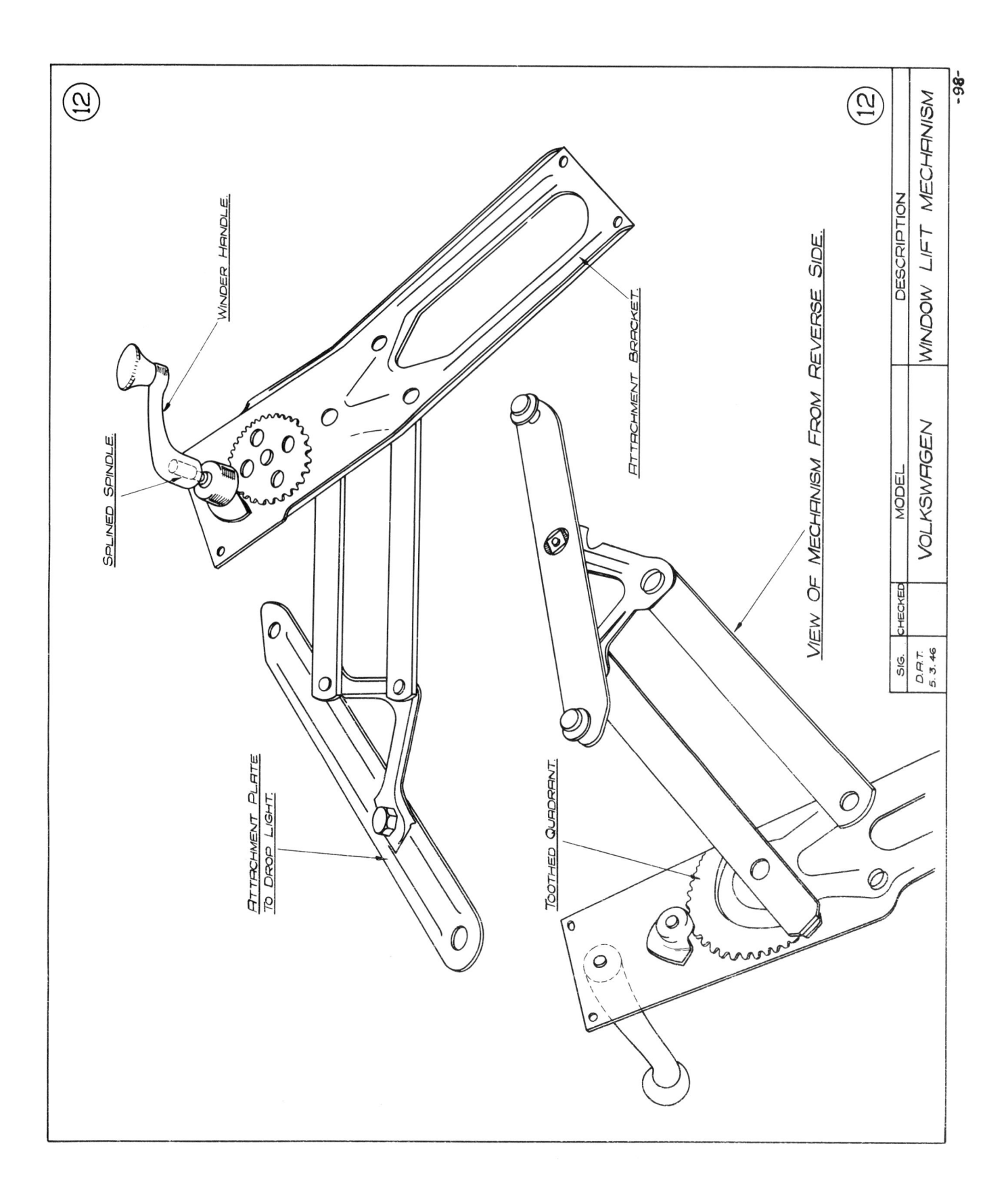

Model: Volkswagen

Labels:
- Winder Handle
- Splined Spindle
- Attachment Bracket
- Attachment Plate to Drop Light
- Toothed Quadrant

View of Mechanism from Reverse Side.

D.R.T. 5.3.46

HANDLE TURNED IN
DIRECTION SHEWN
TO LOCK.

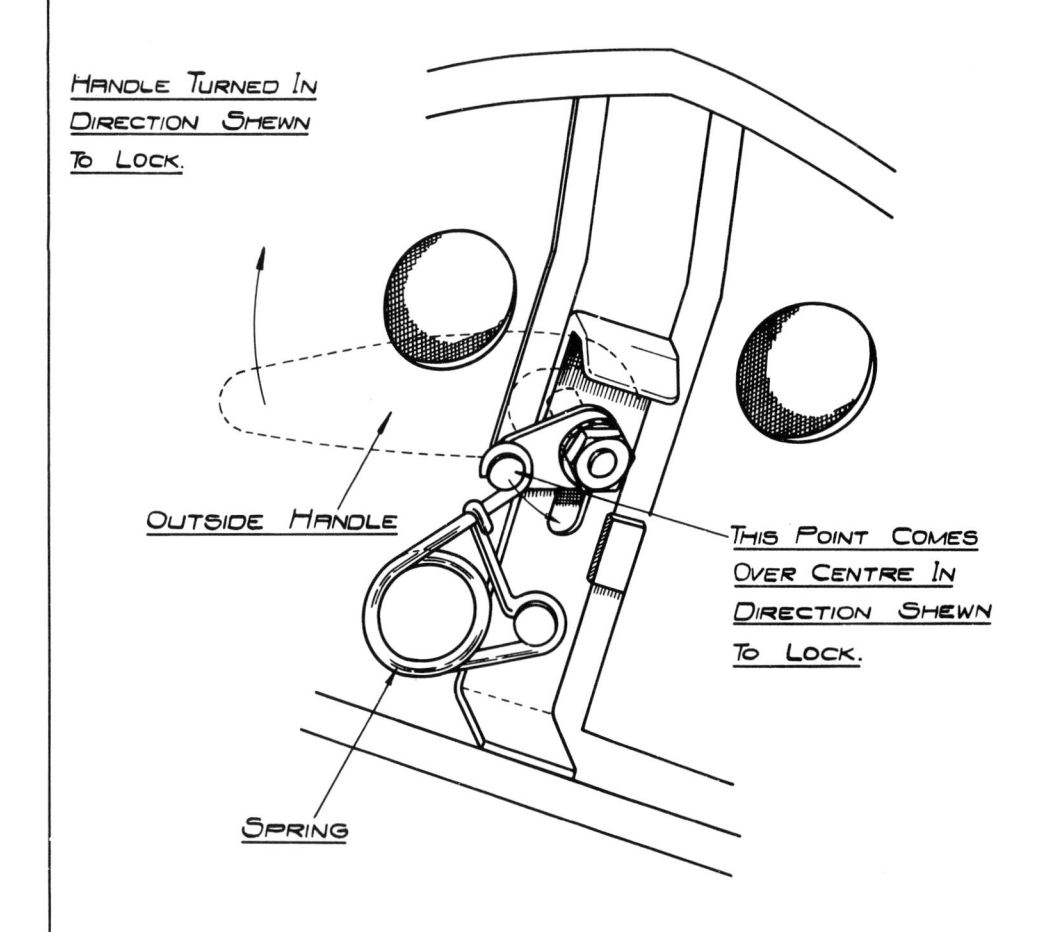

OUTSIDE HANDLE

THIS POINT COMES
OVER CENTRE IN
DIRECTION SHEWN
TO LOCK.

SPRING

SIG.	CHECKED	MODEL	DESCRIPTION
D.A.T. 1.3.46		VOLKSWAGEN	BONNET & ENGINE COMP^T CATCH

(14)

POSITION OF ARMS DURING RAISING OF LID.

POSITION OF ARMS IMMEDIATELY BEFORE LOCKING.

ARM ATTACHED TO BONNET LID.

SUPPORT IN LOCKED POSITION.

ARM ATTACHED TO BONNET SIDE.

GUIDE.

VIEW FROM REVERSE.

(14)

SIG.	CHECKED	MODEL	DESCRIPTION
D.A.T. 1·3·46		VOLKSWAGEN	BONNET SUPPORT MECHANISM

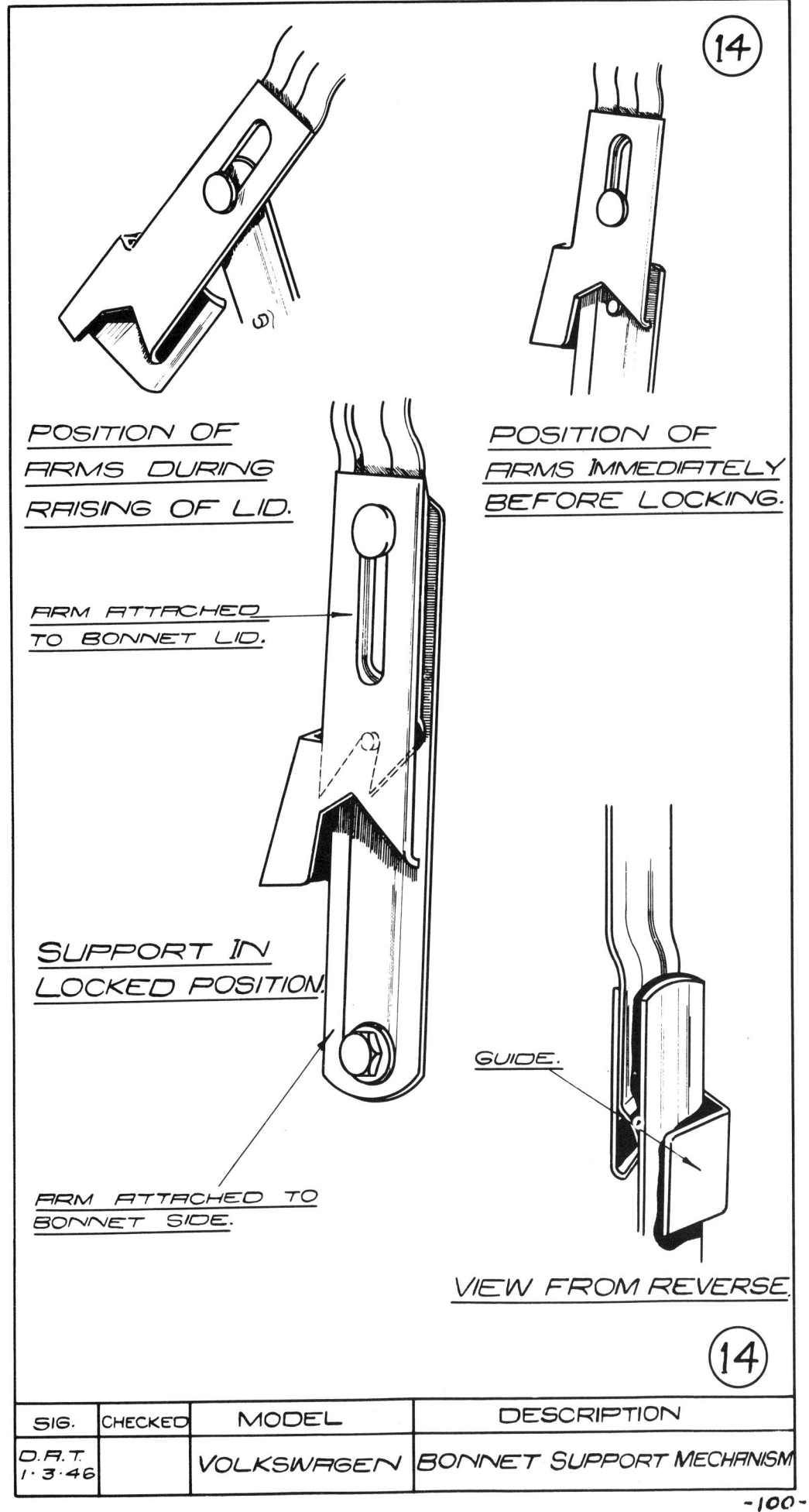

⑮

ATTACHMENT OF HINGE TO BODY

HINGE PIN

ENGINE COMP<u>T</u> LID

LID RETAINED IN RAISED OR LOWERED POSITION BY SPRING END BEING THROWN OVER CENTRE

⑮

SIG.	CHECKED	MODEL	DESCRIPTION
O.R.T. 5·3·46		VOLKSWAGEN	ENGINE COMP<u>T</u> HINGE

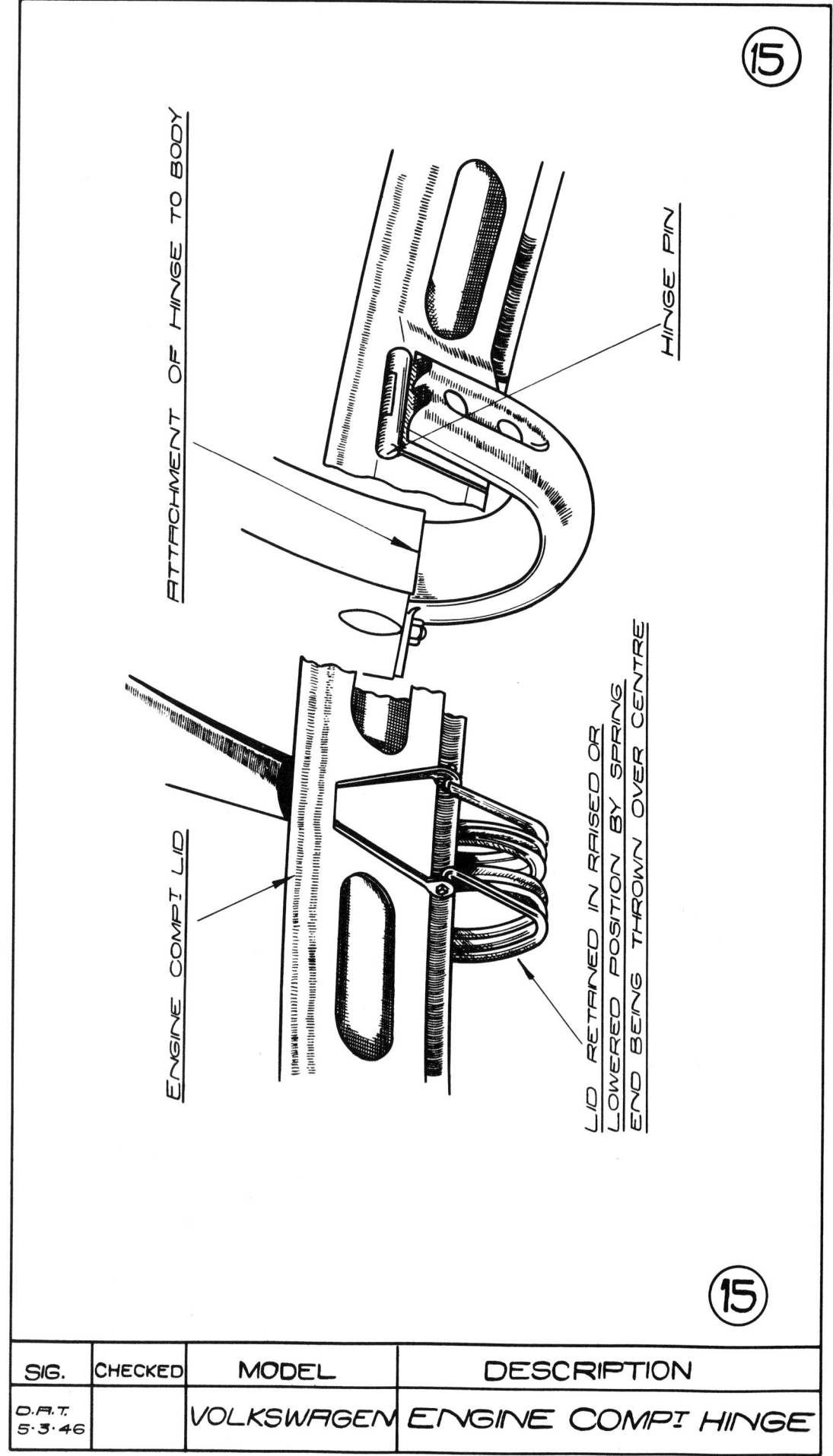

SIG.	CHECKED	MODEL	DESCRIPTION
D.A.T. 7.3.46		VOLKSWAGEN	REAR SEAT CONSTRUCTION

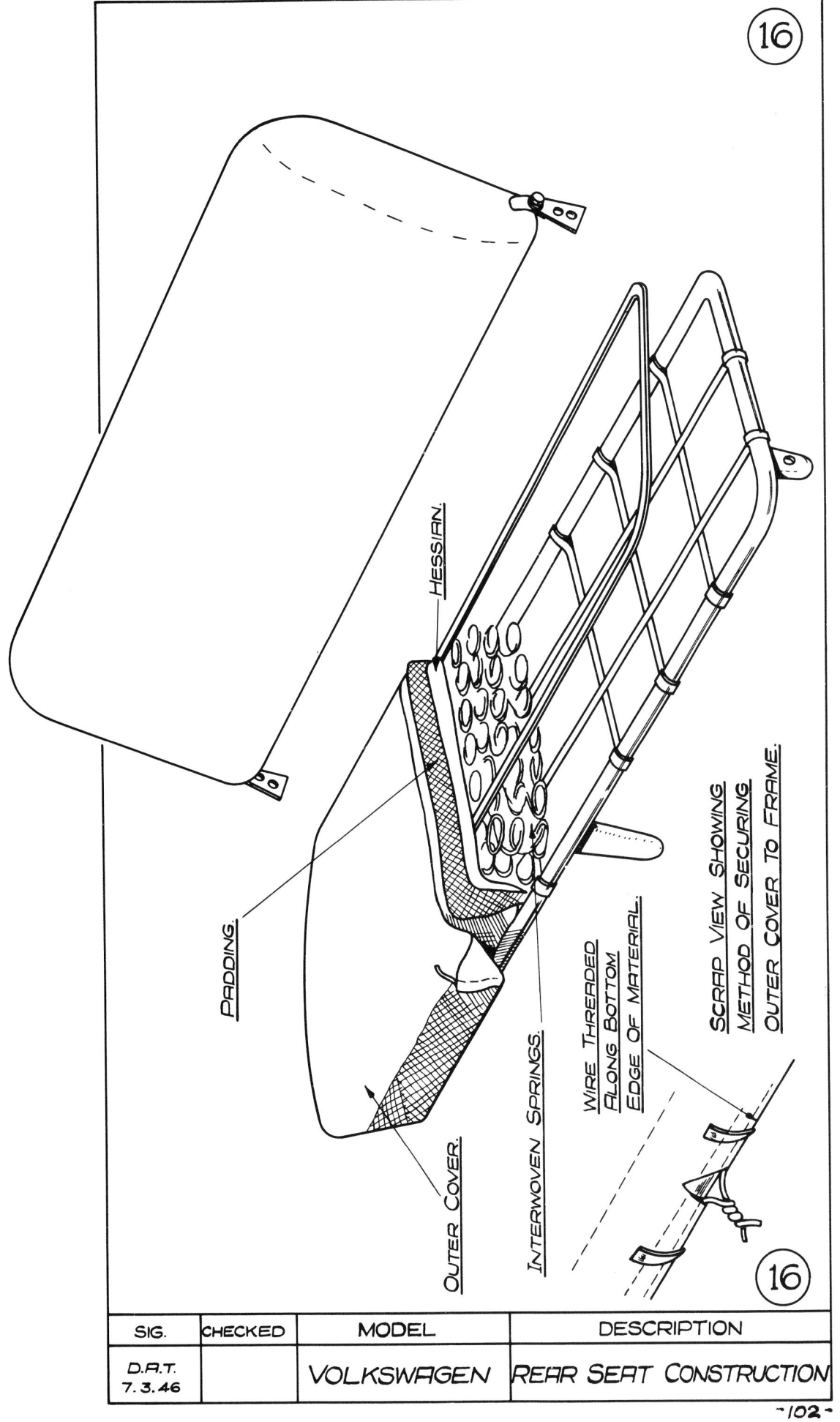

- ENGINE FAN
- WARM AIR DIFFUSED OVER SCREEN FOR DEMISTING.
- BUTTERFLY VALVE
- REMOTE CONTROL (OPERATING VALVE AT REAR)
- WARM AIR OUTLET FOR PASSENGER COMPARTMENT.
- WARM AIR OUTLET FOR DRIVER'S COMPARTMENT.

THIS DRAWING IS DIAGRAMMATIC

SIG.	CHECKED	MODEL	DESCRIPTION
DAT. 12·3·46		VOLKSWAGEN	HEATING SYSTEM.

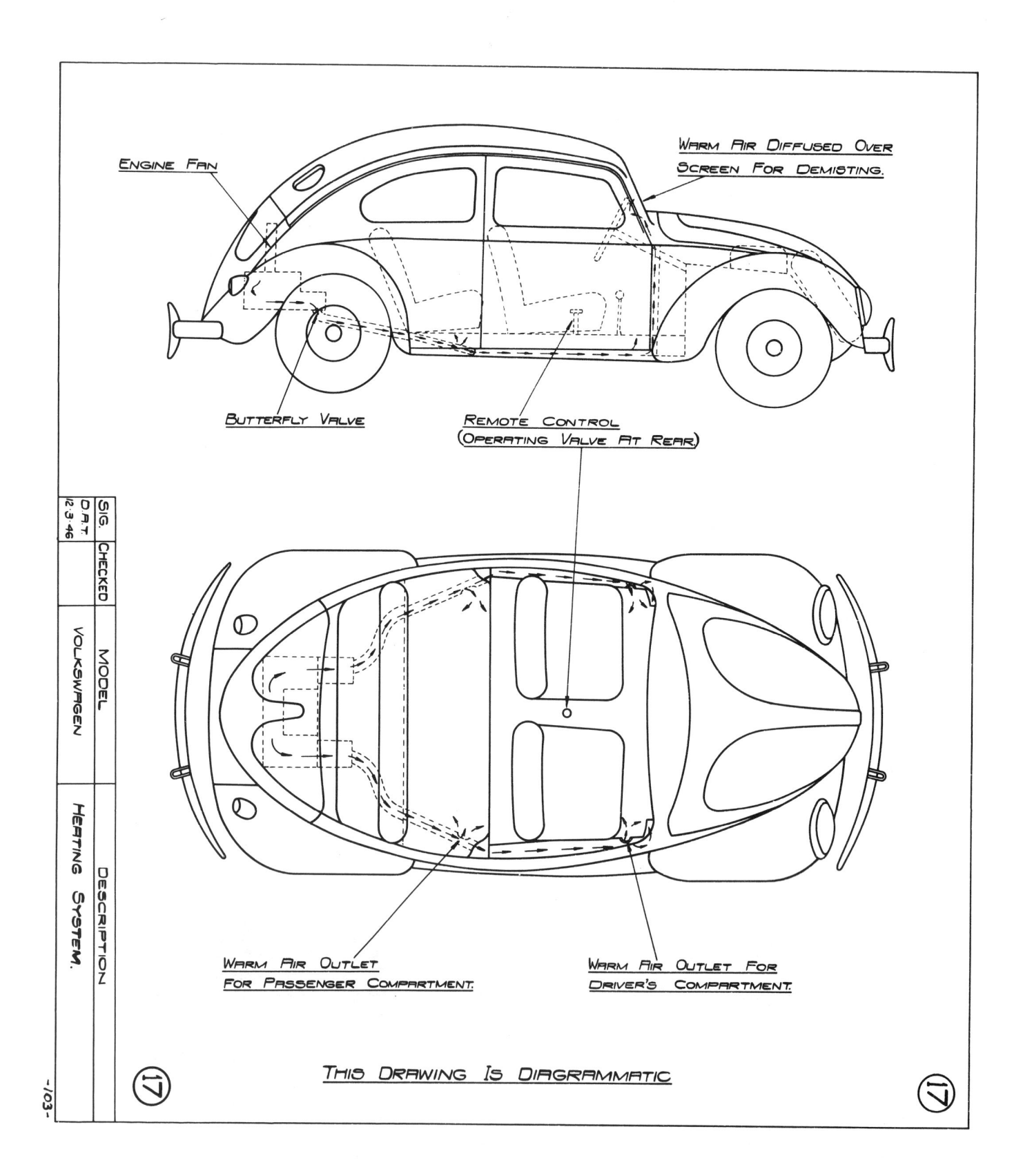

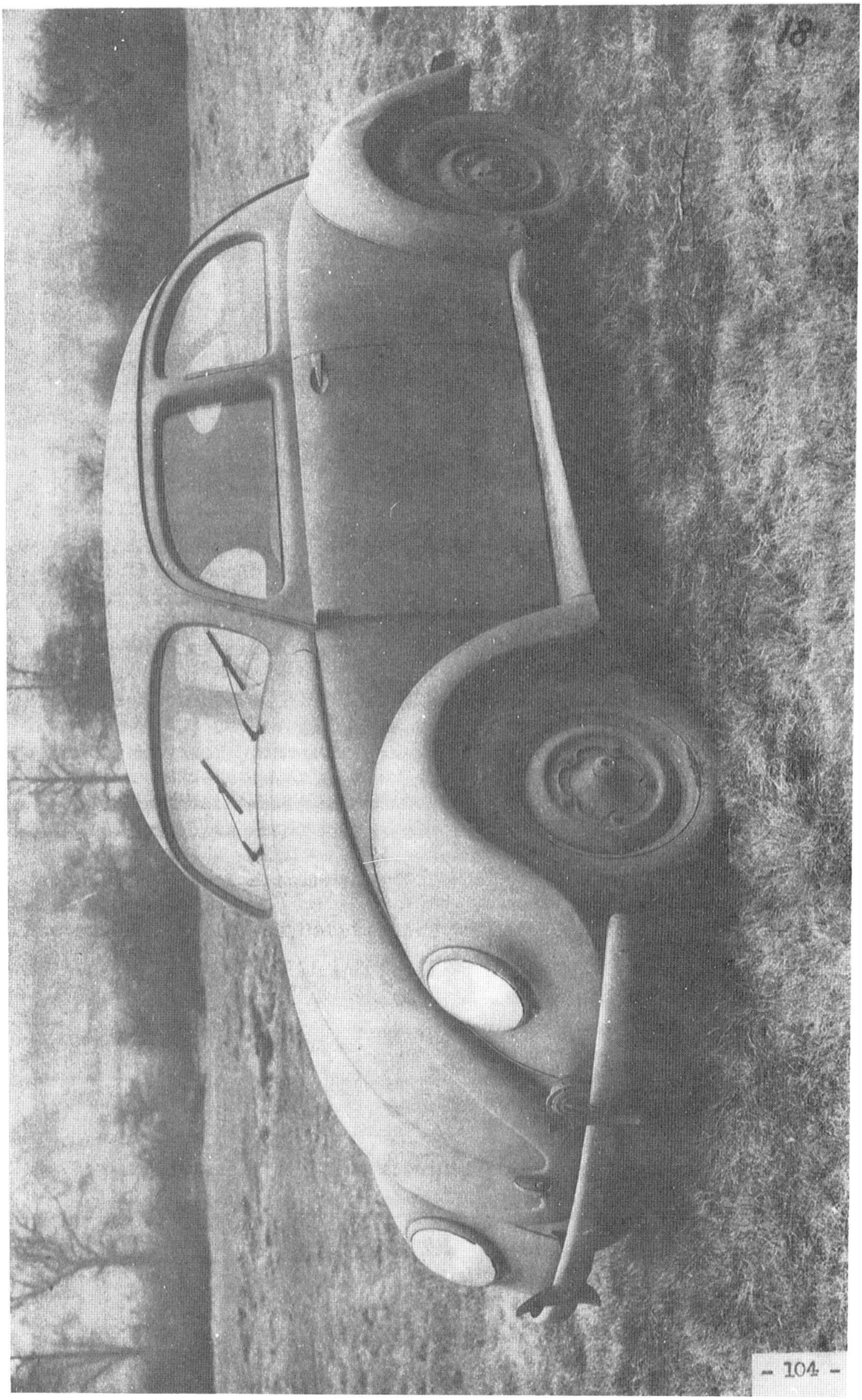

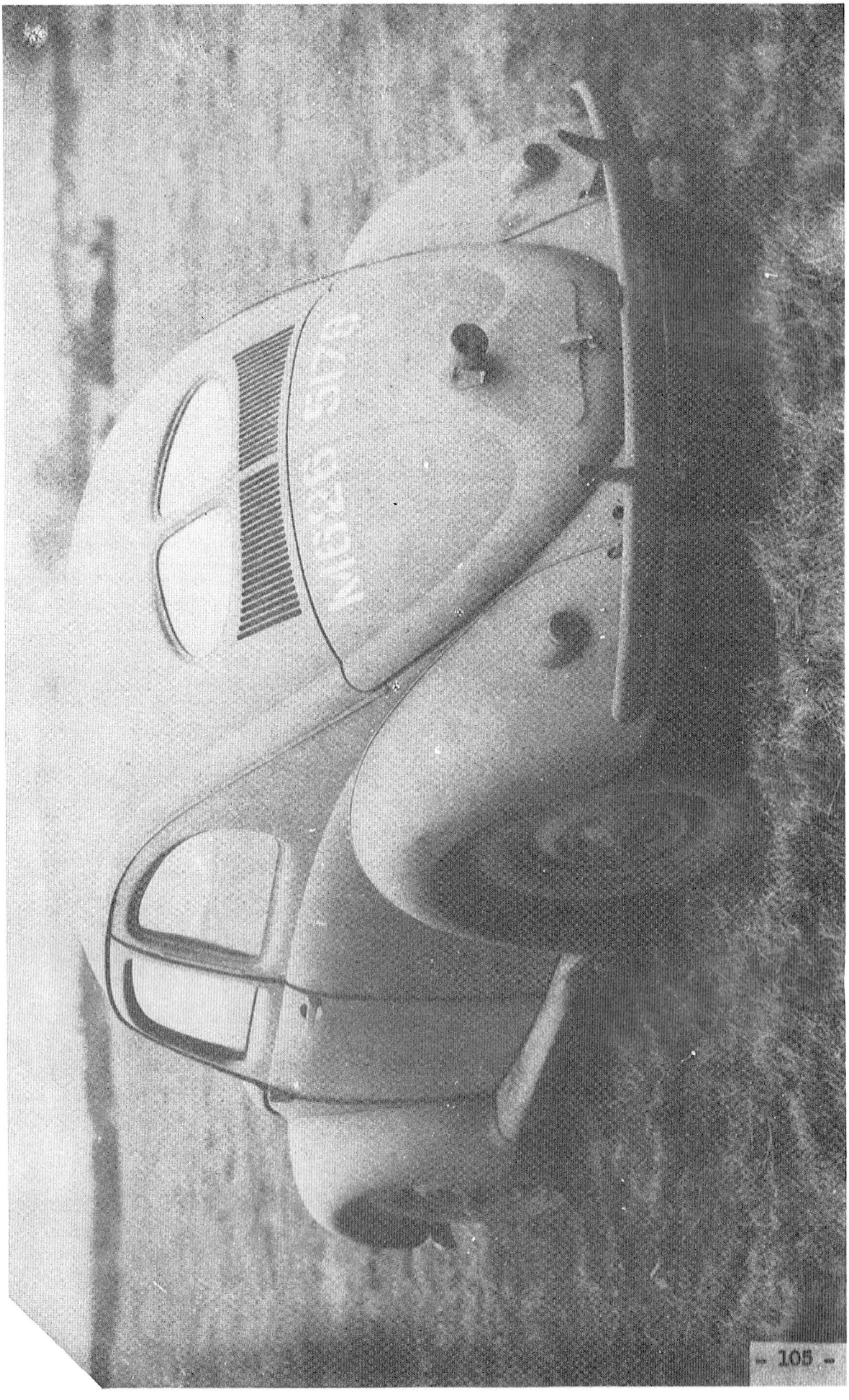

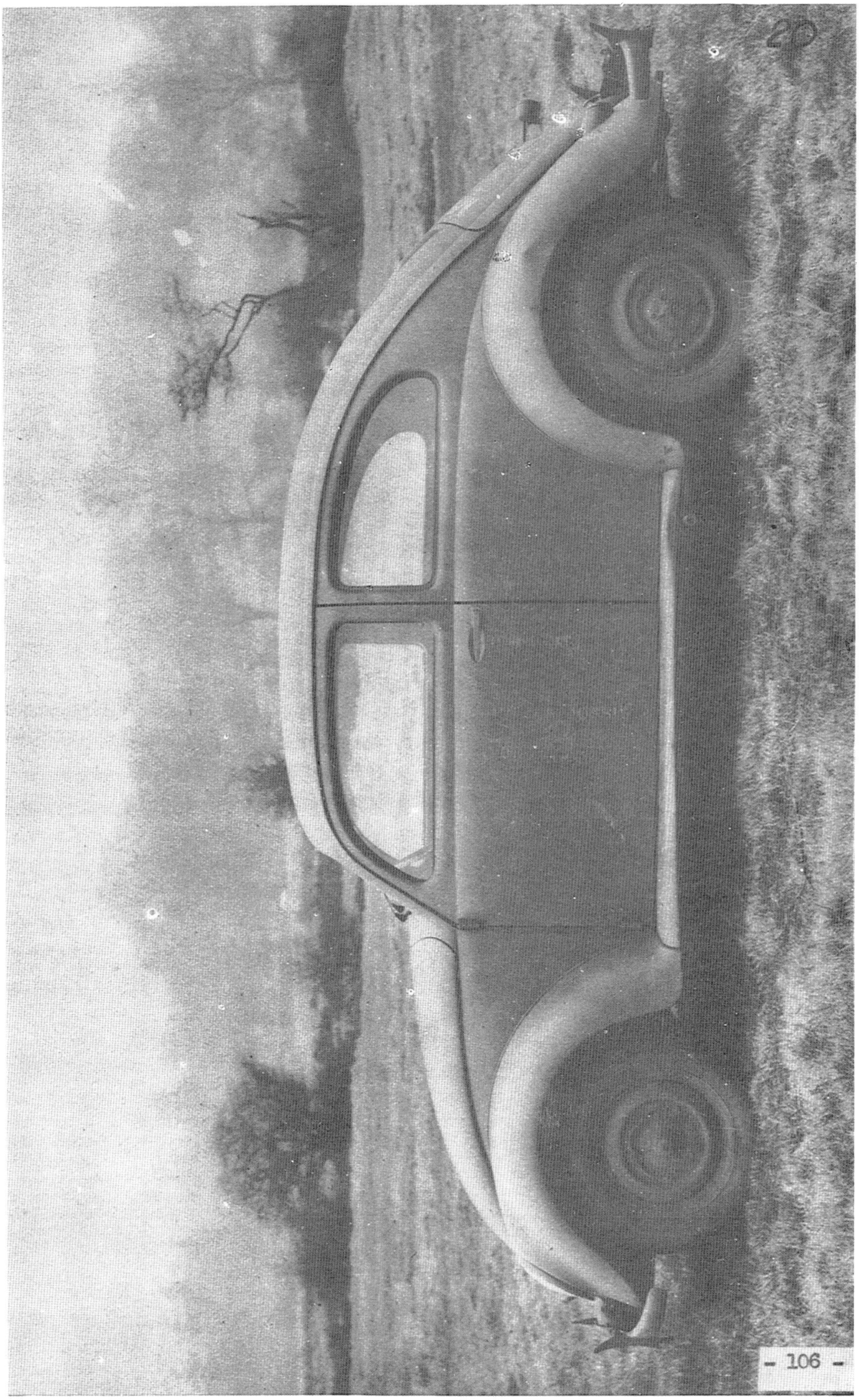

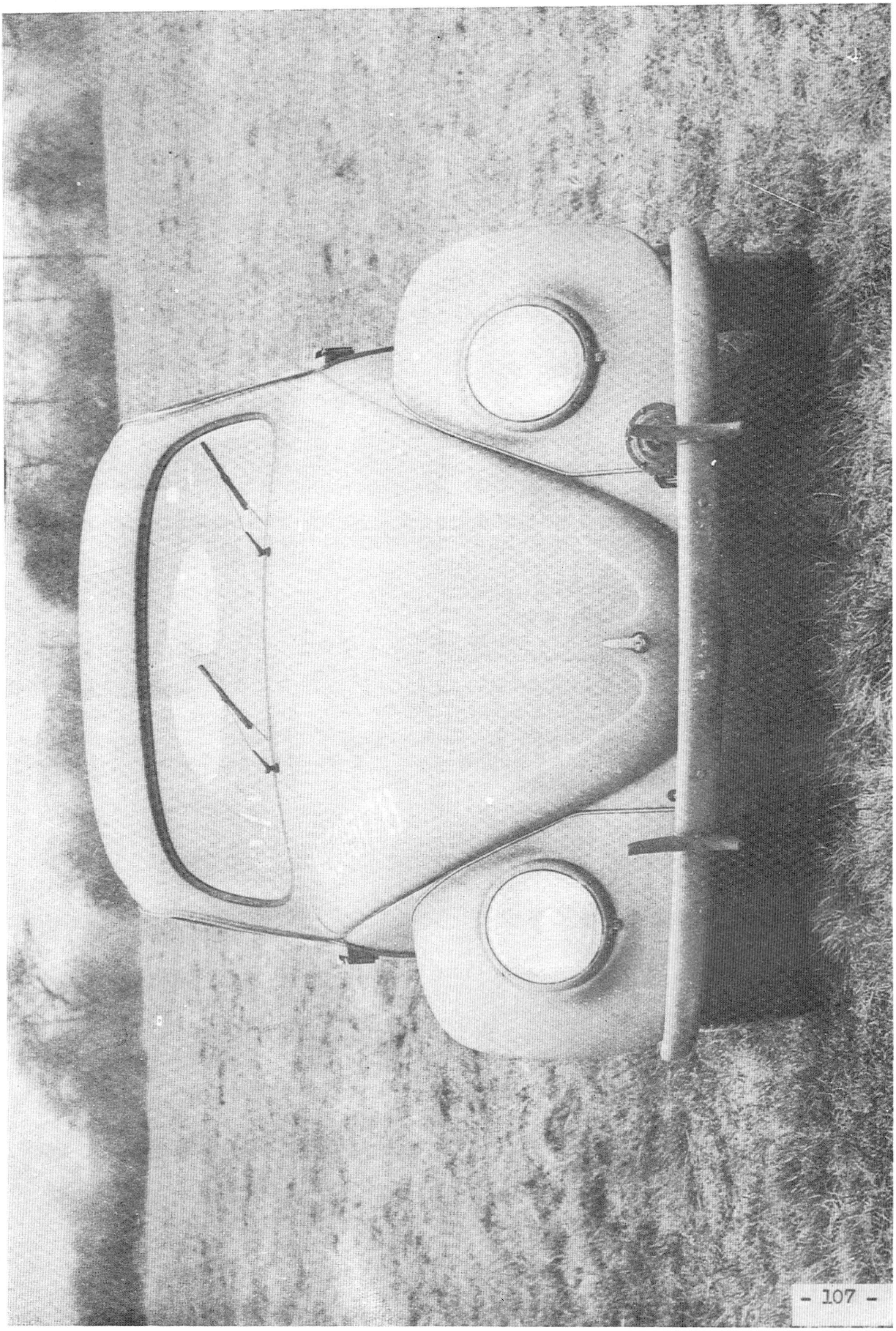

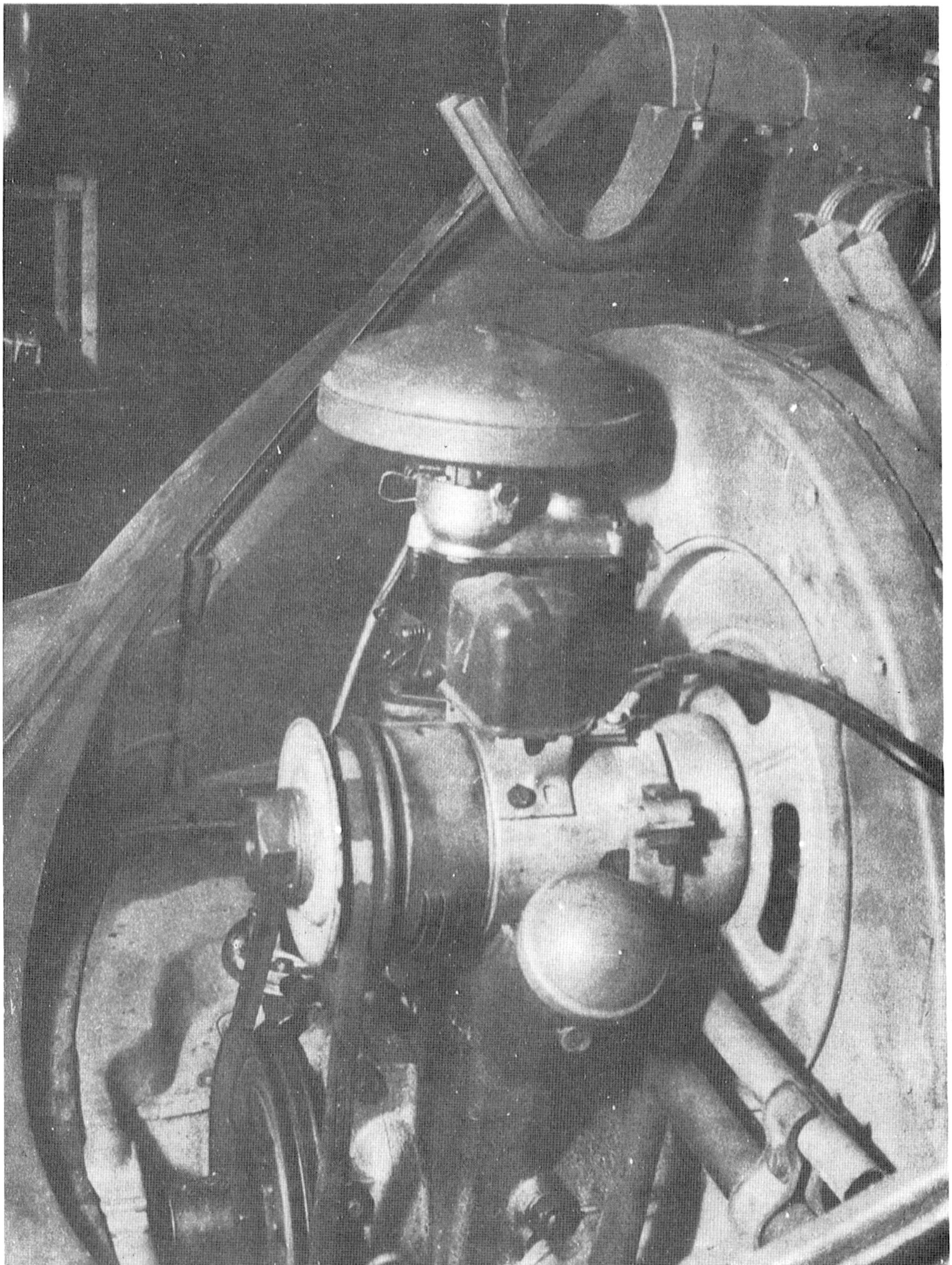

- 109 -

- 110 -

- 111 -

- 112 -

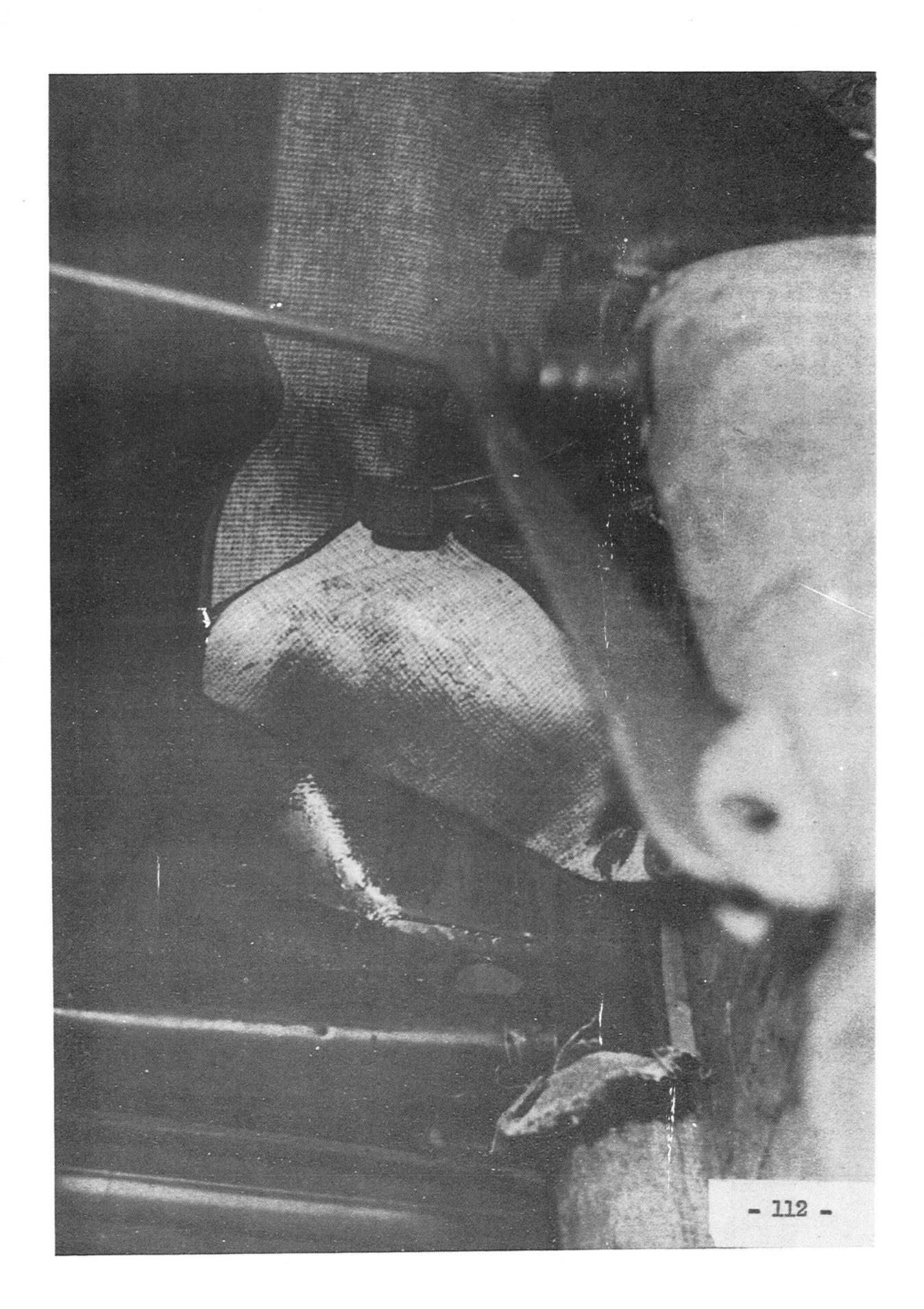

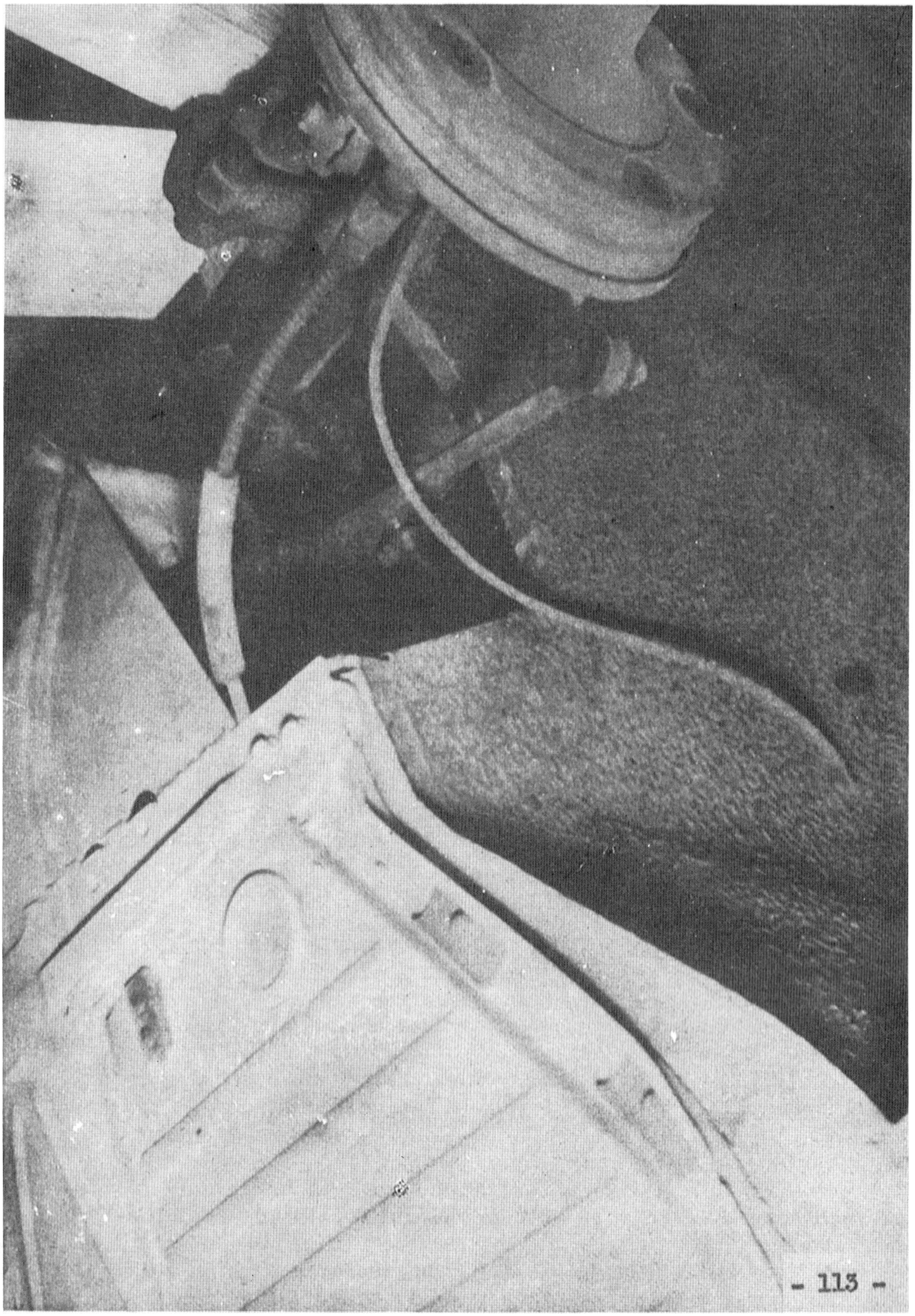

PART IV.

ROAD PERFORMANCE TEST OF THE VOLKSWAGEN MILITARY

VEHICLE, TYPE 21.

Report by:-   SINGER MOTORS LTD.

The vehicle was collected from London and with but a cursory inspection driven straight to Birmingham. Since it was understood that only two miles had been covered prior to this run, the car was not called upon to show its true capabilities. It was, however, apparent from the start that the top gear performance up to 50 m.p.h. and on hill climbing was extremely good, but that this performance was obtained by the use of low overall top gear ratios (overall 6.2:1) and at the expense of the performance above 50 m.p.h.

The preliminary inspection showed the front N.S.Stub axle bearings to be collapsed, that no speedometer cable was fitted, that the horn push on the steering head was broken, that there were severe dents in the wings, that the screen wiper was not working and that the engine compartment lid stay had broken off from the lid.

In spite of the sloppy wheel bearings, steering was positive at all speeds, the defect being only noticeable at about walking pace.

The suspension, if slightly hard, was nevertheless good under all road/speed conditions, both pitch and roll being negligible.

The road holding, as would be expected with the wide track and stiff suspension was extremely good, this quality possibly being accentuated on left hand bends by the left hand steering position.

The brakes were poor and had a slight tendency to pull to the left - extreme newness may have been partly responsible.

The transmission was extremely noisy and very harsh at all speeds in all gears, the differential being the worst item, on sharp bends this noise became intolerable, the final drives had a bad period at approximately 35 m.p.h.

Clutch both smooth and light in operation. Straight through gear box was very pleasant to handle.

Both engine and cooling fan noise was excessive under all conditions - tappets, valve gear etc.

As a purely military general duty or L.A.D. vehicle there is little criticism which can be offered against the general performance and characteristics of the type V.W.82"Volkswagen". The brakes were certainly poor, but this may not be a general fault. The low overall gear ratio is intentional, bearing in mind the type of terrain the car is likely to encounter, the final gear reduction being a refinement, or otherwise, fitted only to the military edition. Excessive noise has never been a factor which has received attention in military vehicles.

The body was removed from the chassis and the engine stripped for investigation. Prior to this, vehicle had been road tested and the following figures obtained:-

Braking from 30 m.p.h.   44 ft. equals 22 ft/sec.^{2} equals 68.5% efficient.

ACCELERATION:

2nd - From standing start 175 ft. in 6 4/5 sec.
     i.e. 0-35 m.p.h. in 6 4/5 sec. in Second Gear.

3rd - From standing start 400 ft. in 13.2/5 sec.
     i.e. 0-41 m.p.h. in 13 2/5 sec. in Third Gear.

Top - From standing start (just rolling) 1157 ft. in 28 2/5 sec.
     i.e. 0-55 m.p.h. in 28 2/5 sec in Top.

        Maximum 1 mile in 1-min. 10-sec. equals
        51 m.p.h. (Better figure would have been
        obtained over measured 1/4 mile).

Best cruising speed - 40 m.p.h.

Due to lack of speedometer and trip no consumptions were taken.

        Total weight   14 cwts. 1 qr.

        Front Axle weight   5 cwts. 3 qrs. 14-lbs.

        Rear Axle weight   8 cwts. 1 qr. 14-lbs.

After re-assembly the vehicle was driven for 20 miles and performance fixtures checked.

---ooOoo---

PART V.

GENERAL IMPRESSION OF THE VOLKSWAGEN MILITARY VEHICLE,

TYPE 21.

Report by:-   A. C. CARS LTD.

This vehicle has been fully investigated and described by Humber Ltd., from a technical angle, therefore these remarks are confined to general handling, characteristics, etc.

1.   The first impression gained in starting up is the extreme noise of the engine, this is immediately blamed on to the cooling fan, which however, is not responsible for more than 10% of the noise, tests carried out with the fan belt removed proved this point.

Examination of the rear end of the vehicle shows that no attempt has been made to reduce engine noise, in fact the design is such that the whole bodywork is acting as a diaphram and amplifies the normal engine noise.

2.   On moving off from stationary a peculiarity is noticed, one might easily be in trouble if gear changing is carried out with any lock on the steering wheel, this is due to the instant response from the engine, accounted for in our opinion, to the absence of a propeller shaft, together with low gear ratios.

This lack of cushioning between the engine and driven wheels is definitely a feature that one has to be educated to, although this feature is not unpleasant when experience is gained.

3. The road holding quality is excellent due no doubt to all round independent suspension.

The swinging arm suspension for the front wheels is, in our opinion a valuable feature, since gyroscopic forces are eliminated.

The vehicle has the ability to keep a straight course "hands off" even on a comparatively rough surface, although for town work it is not so good, as the steering is definitely heavy for any appreciable lock.

The weight distribution could be improved, although the vehicle in its present form does not show any pronounced tendency to tail heaviness, this is probably the result of a generous track in relation to the wheelbase.

4. The forward position of the driver together with lack of bonnet line makes the aiming of the car difficult but this objection is soon overcome with a little practice, and the excellent forward visibility is then appreciated.

5. Summing up we would state that this vehicle is a useful education for those who think there is a future for the rear engined car, also the case for all round independent suspension is clearly demonstrated to be a valuable feature.

From the general construction one gets the impression that the designer has given just enough but no more, therefore as a war vehicle this is no doubt acceptable, but as a civilian vehicle considerable modification would be required to conform to the standard expected.

-----ooOoo-----

PART VI.

BENCH TEST OF THE ENGINE OF THE VOLKSWAGEN.

Report by:-   FORD MOTOR CO. LTD.

VOLKSWAGEN ENGINE.

PHYSICAL AND PERFORMANCE DATA.

PHYSICAL DATA.	
Number of Cylinders	4
Bore - Inches	2.954
" Millimetres	75.03
Stroke - Inches	2.520
" Millimetres	64.0
Capacity Cu. Inches	69.08
" Cu. Cms.	1132
RAC Rating - HP -	13.95
Mean Piston Speed at 1000 RPM (Ft.per.min)	420
Valve Arrangement	O.H. Push Rod
Connecting Rod to Crank Ratio	4.04
Stroke to Bore Ratio	0.854
Compression Ratio	5.63
Weight - Lbs.	189

PERFORMANCE DATA.	
Maximum B.H.P.	24.3
at R.P.M.	3250
Max Torque Lbs/ft	51.3
Max BMEP Lbs.Sq.In	112
@ R.P.M.	2000
B.H.P. Per Litre	21.7
Min.Fuel Consumption Lb/Bhp/Hr.	0.67
at R.P.M. Full Load	2250
Fuel Consump. at Max Torque Lb/Bhp/Hr.	0.69
Average Fuel Consump. Lb/Bhp/Hr. (1000 to 3500 R.P.M)	0.73
Fuel Octane Value	68
Oil	SAE 30
Lbs Weight per BHP.	7.77

Weight is taken with Dry Engine, and includes Silencer and Exhaust System, Car Interior Heaters, and Steel Valances.

VOLKSWAGEN ENGINE.

Performance Record on Dynamometer Test.

---oOo---

The engine was mounted on the dynamometer by 2 brackets at the flywheel end picking up the holes used to bolt the engine to the transmission, and by a single point mounting at the other end, using a bolt which joins the two halves of the crankcase.

Temperatures were recorded as follows:

1. Inlet manifold.
2. Carburettor air inlet.
3. Sump oil.
4. Inlet cooling air.
5. Outlet air temperature on the right bank.
6. Outlet air temperature on the left bank.

Mercury manometer records were taken as under:-

1. Inlet manifold, in riser just after carburettor.
2. In exhaust pipe of No.2. cylinder before the silencer.

The engine was run in at varying speeds up to 3000 r.p.m. full load, maximum torque being periodically recorded. After 10 hours running no further increase in torque was observed. The following measurements were then taken and the valve clearances, contact breaker and spark plug gaps set as under. After which the performance tests recorded were taken.

```
Cylinder bore dia........................2.954 ins.
Stroke..................................2.52  ins.
Total swept vol........................1132 ccs.
Total clearance vol....................244 ccs.
Compression ratio......................5.63 : 1
Valve clearances set (inlet & exhaust).... .005 ins.
Distributer contact breaker gap........... .012 ins.
Spark plugs gaps.......................... .025 ins.
```

Solex 26 V.F.1. Carburettor setting.

```
Throttle tube........................26 m.m.
Choke...............................21.5
Main jet............................120
Air correction jet..................170
Pilot............................... 45
Emulsion tube.......................100
Dynamometer.........................Heenan and Froude DPX.3.
Fuel................................MT.68 sp. gr. .740.
Engine oil..........................SAE.30.
```

TEST 1.   Full load curve.

Before the tests were commenced the distributor was set to give optimum power at 1750 r.p.m.   This setting was not altered through the test.

Attempts to run at steady speeds between 2400 and 2800 and 2800 r.p.m. set up violent vibrations of the inlet manifold and carburetter assembly resulting in speed fluctuations of as much as 500 r.p.m.   The vibrations decreased steadily above this latter speed and the engine was again steady at 3500 r.p.m.   In order to complete the readings at the above speeds the carburettor was steadied by a packing between it and the main cooling air duct.

The difference in the temperatures of the outlet air of the right bank as compared with that of the left bank appears to be an unsatisfactory feature.

The specific fuel consumption was high, a minimum of .670 lbs/bhp/hr occurring at 2000 rpm.   Exhaust gas analysis taken at 1500 rpm and 2000 rpm confirmed that the mixture was rich, 11.61 : 1 and 12.93 : 1 respectively, being air/fuel ratio at these speeds. In view of the high consumption, a spot reading was taken at 1750 rpm with the carburettor setting changed:-

	Torque.	Specific Fuel Consumption.
	lbs/ft.	lbs/bhp/hr.
Std. carburettor setting	51.1	.739
Main jet 115 (replacing 170) Air Corr.(replacing 170) 190	49.6	.619
Percentage difference.	2.9%	16.2%

These readings were taken for information only, as a check on the carburetter of another engine of this type showed that the two settings were the same.   All tests recorded were, therefore, made with the carburettor setting as received.

TESTS 2, 3, and 4.
$\frac{3}{4}$, $\frac{1}{2}$ and $\frac{1}{4}$ load curves.   (See graphs 1 to 4).

The vibration and the difference of temperature of outlet cooling air of the two banks noted in Test 1 apply in Tests 2, 3 and 4, these being a function of speed and fan/deflector characteristics respectively.

During Test 3 one of the vertical tubes of the oil cooler burst when the oil pressure was 35 lbs/sq.in. and engine speed was 3000 rpm.

TEST 5.

Morse test with full throttle.
For mechanical efficiency, friction and pumping losses and I.H.P. see graphs 1 and 2.

The difference in the indicated horse power of the different cylinders is probably due to distribution and enhanced by rich mixture strength.

GENERAL.

Some pinking was observed at speed range of 1000 rpm to 2000 rpm.
There was considerable piston slap when the engine was cold, this was not present when the engine had reached a normal working temperature.

The ducting of the cooling air appears to need some revision to ensure an even heat transference from both banks of cylinders.
The carburation could be improved to give a more economic fuel consumption.
The performance is not outstanding and the general noise in operation is excessive, the engine could not be recommended in its present form.

---ooOoo---

VOLKSWAGEN ENGINE.

Analysis of Exhaust Gases
Recorded by the Orsat Gas Analyser.

PERCENTAGE VOLUME OF CONSTITUENT GASES.

R.P.M.	Gas	Full Load	3/4 Load	1/2 Load	1/4 Load
1500	CO_2	7.2	9.8		
	O_2	0.2	0.2		
	CO	11.2	7.2		
	N_2	81.4	82.8		
2000 max torque	CO_2	8.0	10.2	11.0	12.4
	O_2	0.2	0.2	0.2	0.3
	CO	8.4	5.8	5.2	3.5
	N_2	83.4	83.8	83.6	83.8

AIR/FUEL RATIO FROM GAS ANALYSIS.

	Full Load	3/4 Load	1/2 Load	1/4 Load
1500 R.P.M.	11.61 :1	12.76 :1		
2000 R.P.M.	12.93 :1	13.49 :1	13.40 :1	13.81 :1

PERCENTAGE VOLUMETRIC EFFICIENCY FROM A/F RATIO.

	Full Load	3/4 Load	1/2 Load	1/4 Load
1500 R.P.M.	95	69		
2000 R.P.M.	95	73	54	37

FUEL:- White Pool Petrol.   Specific Gravity .740

ESTIMATED FUEL COMPOSITION - H_2 = 14%   C = 86%

	ENGINE UNDER TEST	STANDARD ENGINE.
SPECIFICATION OF VOLKSWAGEN ENGINE	Bore 75.03 mm Stroke 64.0 mm Capacity 1132 cc	Bore 70.0 mm Stroke 64 mm Capacity 985 cc.

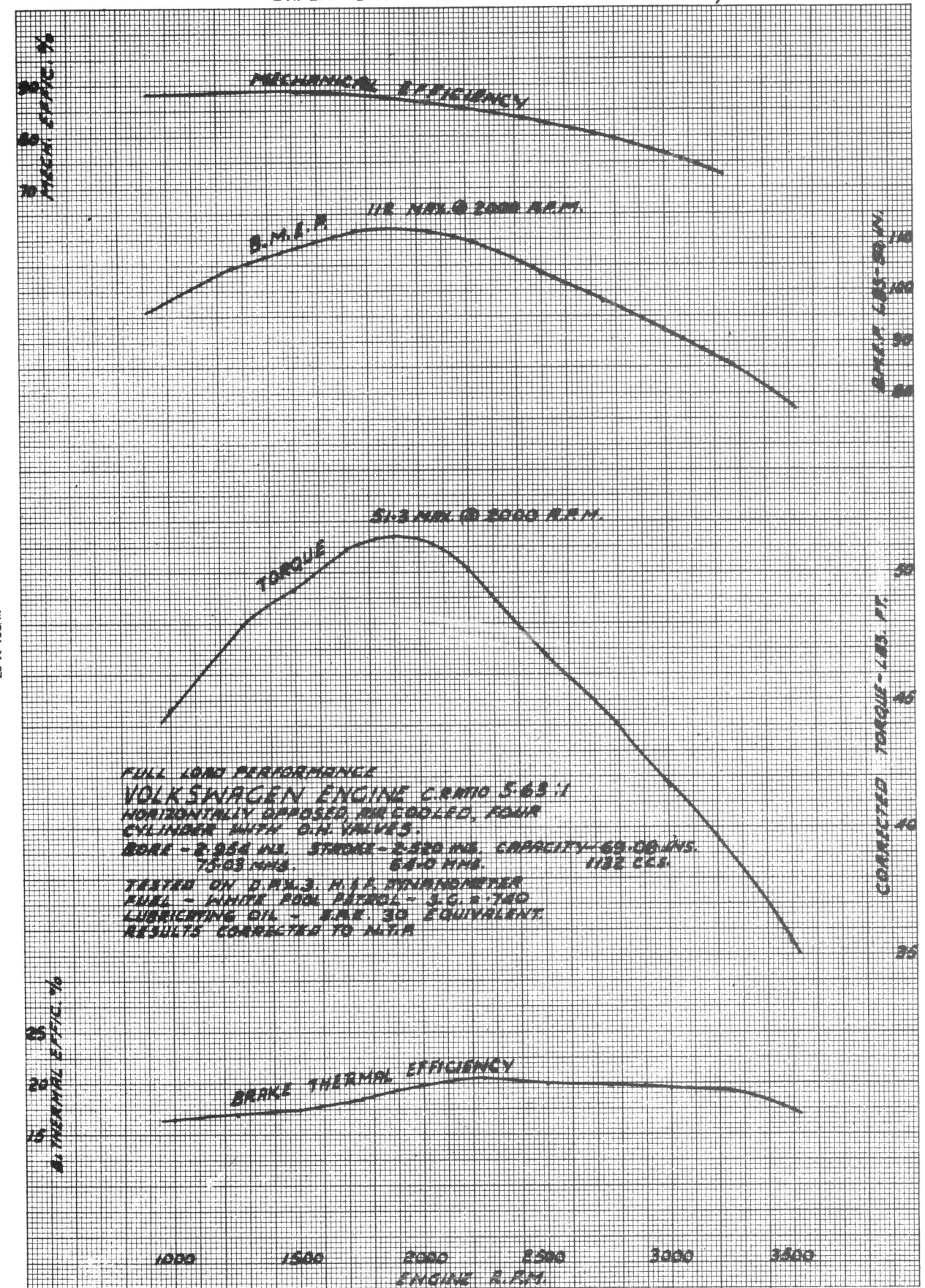

FORD MOTOR COMPANY LTD.
DAGENHAM.
EXPERIMENTAL DEPT.

DATE: FEB. 8 - 1946.
REFERENCE: 2202
GRAPH 1/4.

FULL LOAD PERFORMANCE
VOLKSWAGEN ENGINE C. RATIO 5.63:1
HORIZONTALLY OPPOSED, AIR COOLED, FOUR
CYLINDERS WITH O.H. VALVES.
BORE - 2.954 INS.   STROKE - 2.520 INS.   CAPACITY - 69.06 INS.
       75.03 MMS.   64.0 MMS.   1132 C.C.S.
TESTED ON D.P.W.3. P.I.P. DYNAMOMETER
FUEL - WHITE POOL PETROL - S.G. = .740
LUBRICATING OIL - S.A.E. 30 EQUIVALENT.
RESULTS CORRECTED TO N.T.P.

Curves shown: MECHANICAL EFFICIENCY; B.M.E.P. 112 MAX. @ 2000 R.P.M.; TORQUE 51.3 MAX. @ 2000 R.P.M.; BRAKE THERMAL EFFICIENCY.

X-axis: ENGINE R.P.M. (1000–3500)

FORD MOTOR COMPANY LTD.
DAGENHAM.
EXPERIMENTAL DEPT.

DATE FEB. 8. 1946
REFERENCE 2202
GRAPH. 2/4.

FULL LOAD PERFORMANCE
VOLKSWAGEN ENGINE
FOR DETAILS SEE GRAPH 1

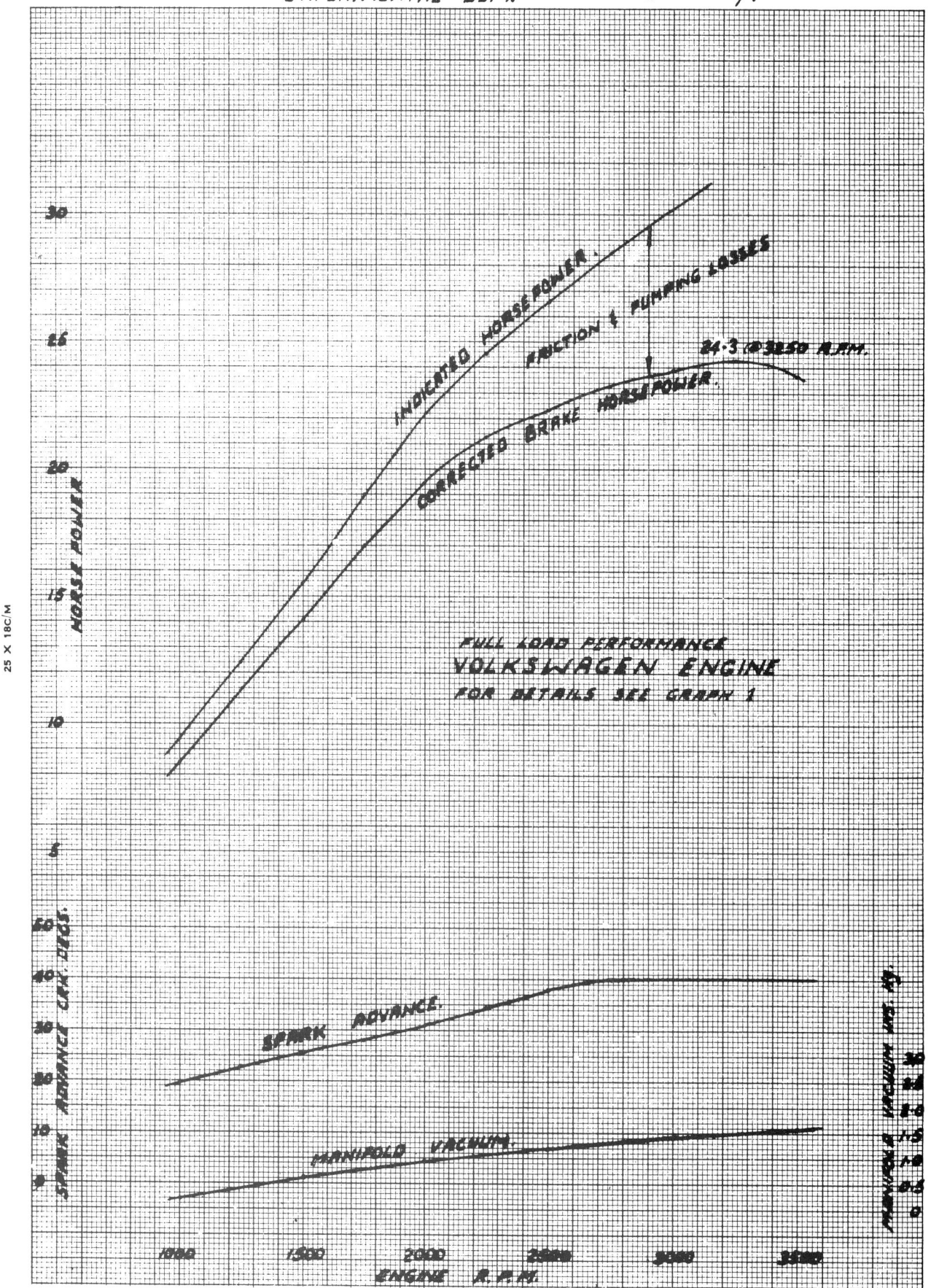

INDICATED HORSE POWER.
FRICTION & PUMPING LOSSES
CORRECTED BRAKE HORSE POWER.
24·3 @ 3250 R.P.M.

SPARK ADVANCE.
MANIFOLD VACUUM.

HORSE POWER
SPARK ADVANCE CRK. DEGS.
MANIFOLD VACUUM INS. Hg.
ENGINE R.P.M.

FORD MOTOR COMPANY LTD.
DAGENHAM.

EXPERIMENTAL DEPT.

DATE FEB. 8. 1946
REFERENCE 2202
GRAPH 3/4

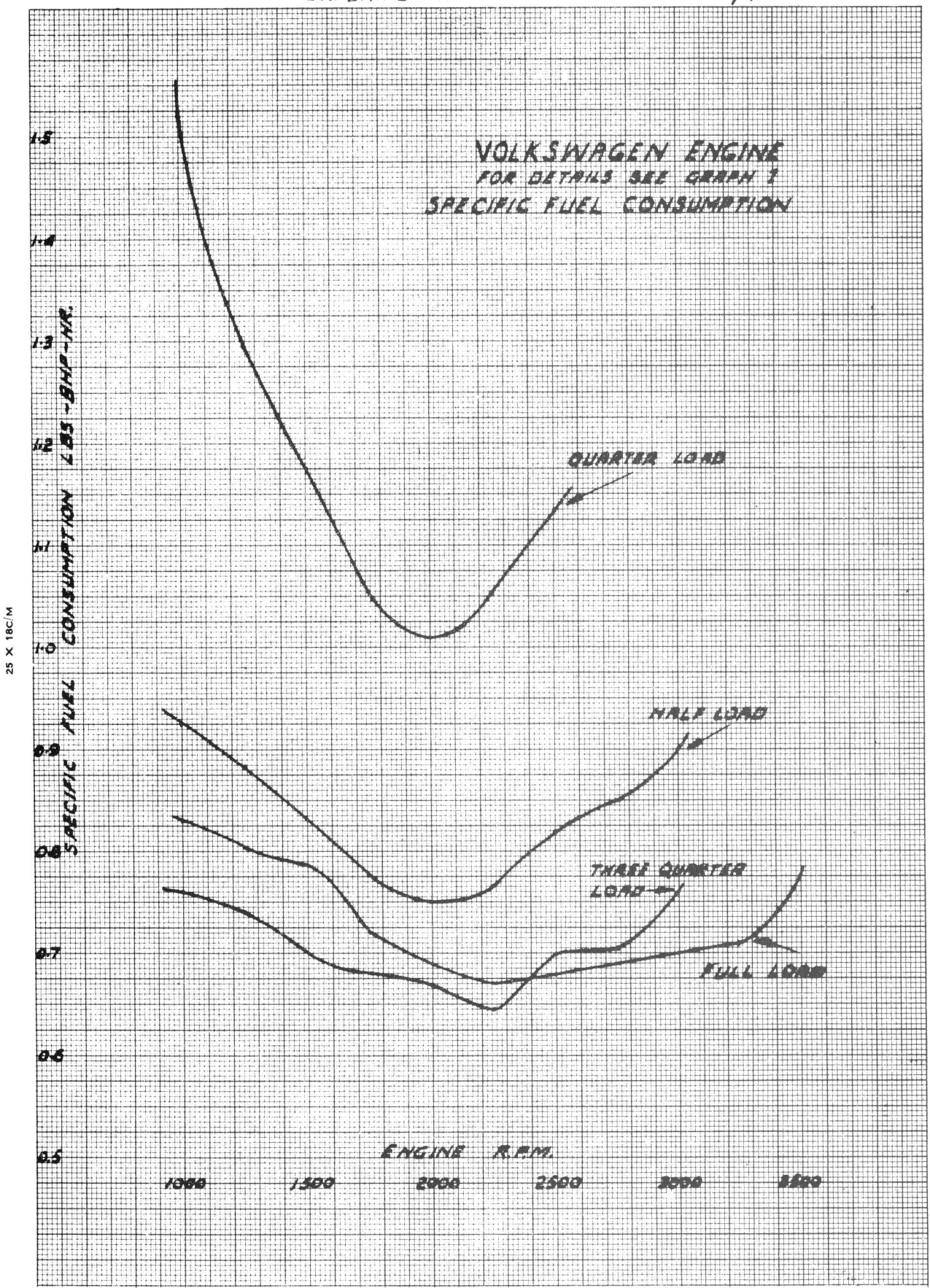

VOLKSWAGEN ENGINE
FOR DETAILS SEE GRAPH 1
SPECIFIC FUEL CONSUMPTION

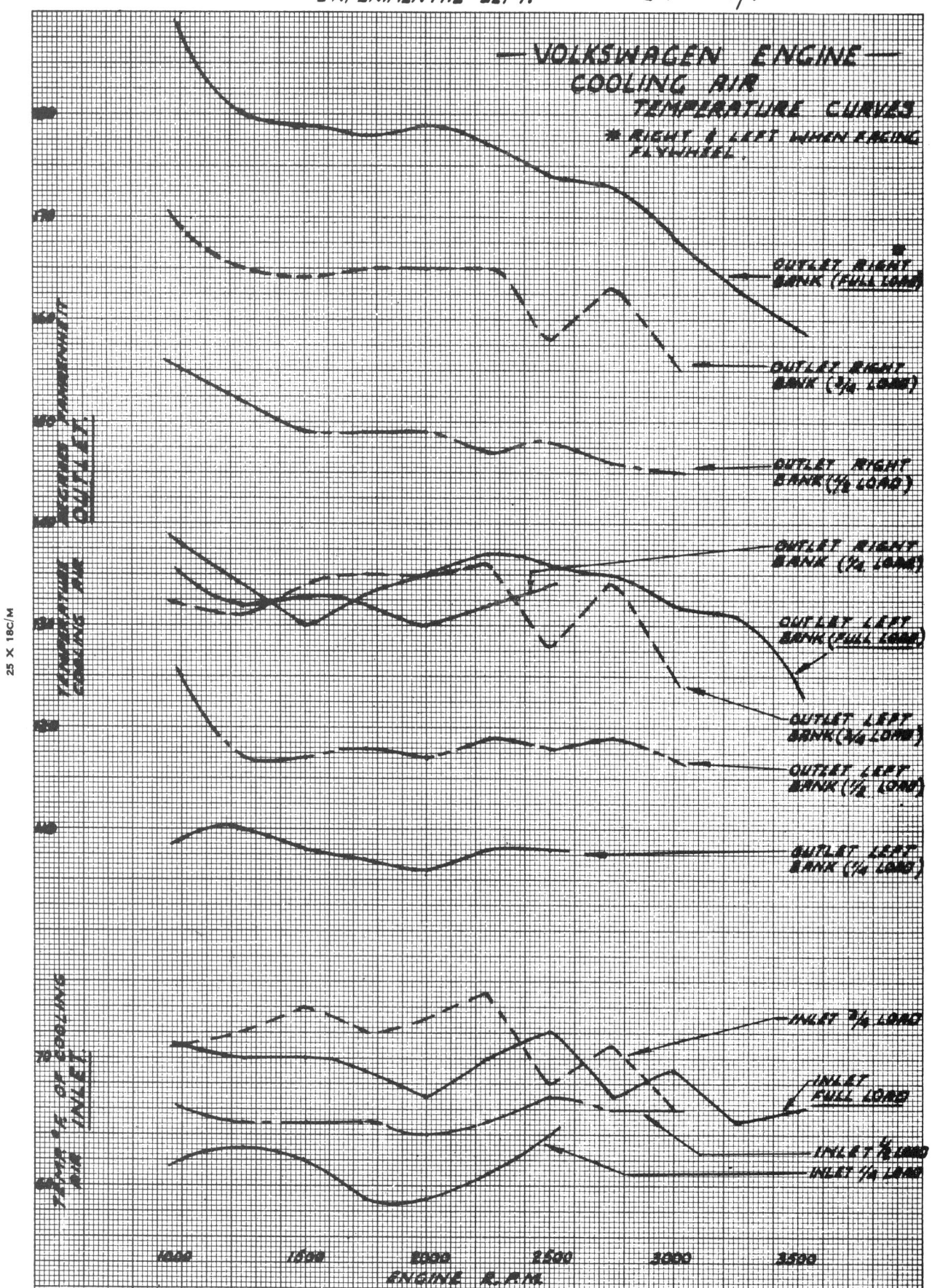

FORD MOTOR COMPANY LTD.
DAGENHAM.
EXPERIMENTAL DEPT.

DATE FEB. 8TH 1946
REFERENCE 2202
GRAPH 4/4

VOLKSWAGEN ENGINE COOLING AIR TEMPERATURE CURVES

* RIGHT & LEFT WHEN FACING FLYWHEEL.

- OUTLET RIGHT BANK (FULL LOAD)
- OUTLET RIGHT BANK (3/4 LOAD)
- OUTLET RIGHT BANK (1/2 LOAD)
- OUTLET RIGHT BANK (1/4 LOAD)
- OUTLET LEFT BANK (FULL LOAD)
- OUTLET LEFT BANK (3/4 LOAD)
- OUTLET LEFT BANK (1/2 LOAD)
- OUTLET LEFT BANK (1/4 LOAD)
- INLET 3/4 LOAD
- INLET FULL LOAD
- INLET 1/2 LOAD
- INLET 1/4 LOAD

ENGINE R.P.M. — 1000, 1500, 2000, 2500, 3000, 3500

-129-

FORD MOTOR COMPANY LTD.
DAGENHAM.
EXPERIMENTAL DEPT.

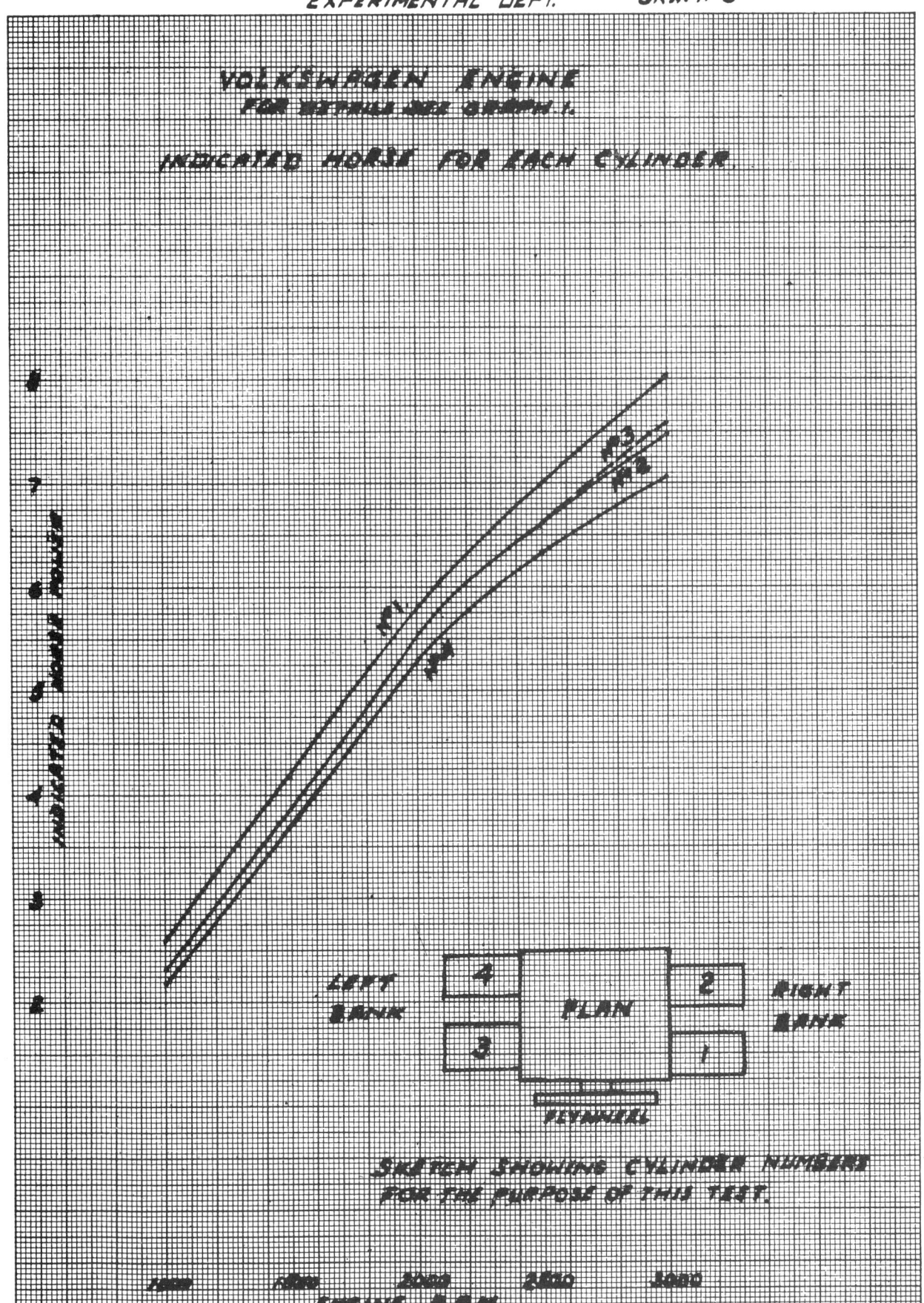

DATE: FEB. 8TH 1946
REFERENCE: 2202
GRAPH 5

-130-

EXP. NO. 2202 - VOLKSWAGEN ENGINE - 1st FEB. 1946
No.1 of 3 views - Full Rear

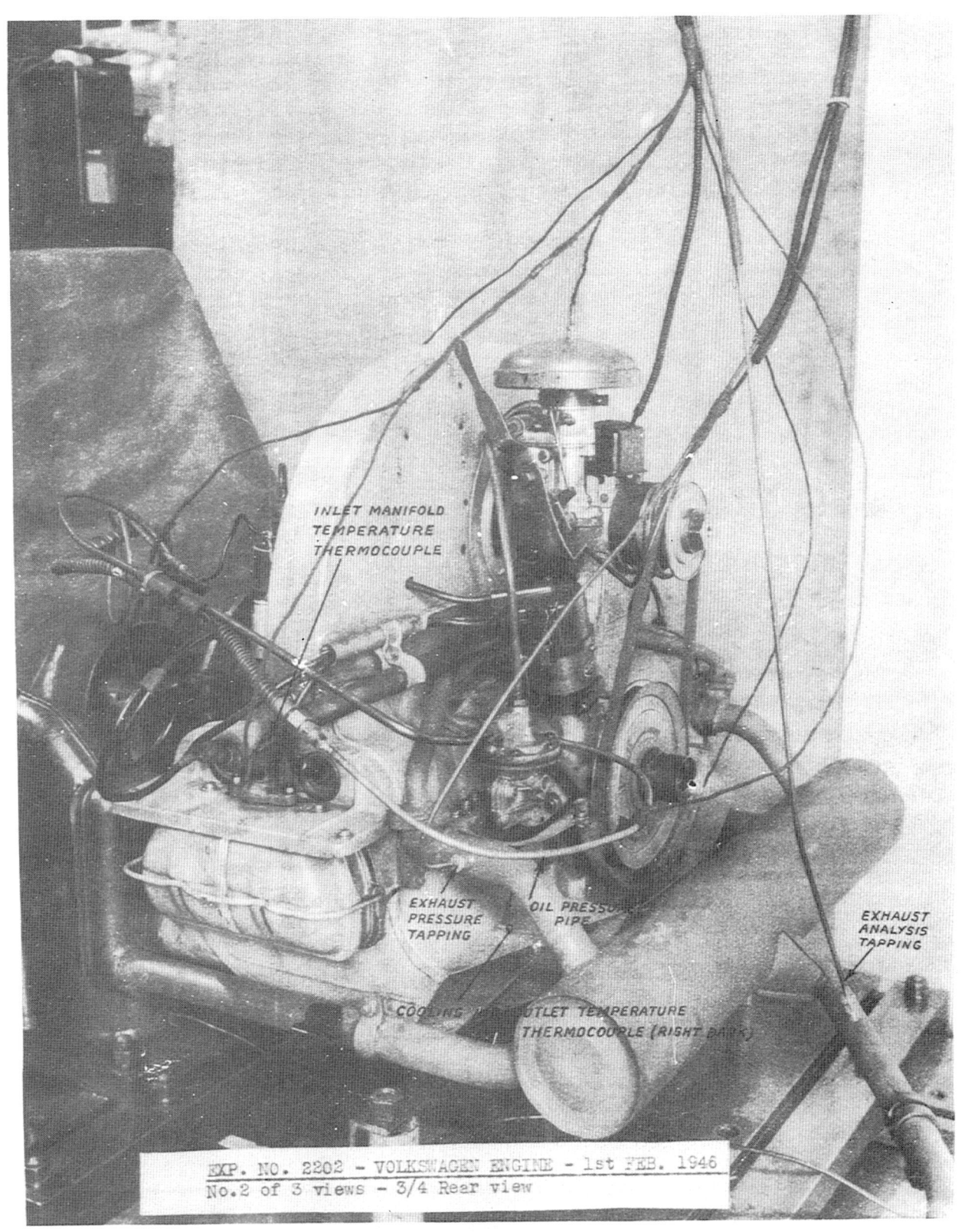

EXP. NO. 2202 - VOLKSWAGEN ENGINE - 1st FEB. 1946
No.2 of 3 views - 3/4 Rear view

COOLING AIR INLET
TEMPERATURE THERMOCOUPLE

SPARK ADVANCE
INDICATOR LEAD

EXP. NO. 2202 - VOLKSWAGEN ENGINE - 1st. FEB 1946
No.3 of 3 views - 3/4 Front View

PART VII

THE CARBURETTOR FITTED TO THE VOLKSWAGEN.

Report by:- SOLEX LTD.

The Solex 26 VF1 carburettor fitted to the Volkswagen has no special features from the carburation angle, and very few ones from the constructional angle.

The jet system is of the usual Solex pattern, called 21 assembly. The jets themselves are interchangeable with those available in this country.

The carburettor has no such refinements as accelerating pump or economy device.

The setting found in the carburettor was choke tube 21.5; main jet 120; correction jet 170; jet well 5.3 m/m; pilot 45; pilot air bleed 1.5; float 12.5 grs; N.V.1.2.

The mixture strength given by the setting is not economical, being about 12.5:1 at full throttle, and 12.0:1 at part throttle.

Starting is obtained by means of a strangler butterfly with spring loaded air valve, and interconnection between the strangler and throttle levers. When the strangler is fully shut, the butterfly is slightly opened by a single link and floating lever arrangement. When the engine fires, the spring loaded valve in the strangler butterfly opens, and thus provides the necessary weight of mixture to keep the engine running.

The carburetter is dustproof. Felt bushes are inserted in the throttle spindle bearings, and a gasket is fitted between the float-chamber and its cover. The float-chamber is, however, unusually dirty inside, due possibly to the petrol not being suitably filtered.

A number of parts usually made of brass over here are in steel in this carburetter, and no attempt appears to have been made to protect them against rust. An exception is the float, which is made of plated steel, but even this shows signs of rust.

Although the carburetter is what is known as a 26 m/m-bore, the flange is that of a 30 m/m carburetter.

The air intake diameter is very substantial, being 46 m/m-bore and 52 m/m outside diameter.

---ooOoo---

APPENDIX.

RULES GOVERNING APPLICATION FOR A VOLKSWAGEN SAVINGS BOOK.

---ooOoo---

(1) The original Volkswagen savings book will be issued by the agent in the applicant's place of residence or work against payment of a fee of 1 mark. The book is made out in the name of the applicant and neither it nor the rights it confers are transferable. Acceptance of the savings book is equivalent to placement of an order for delivery of a Volkswagen, subject to the rules herein set forth.

(2) Current instalment dues are payable at offices of the labour Front or the Strength-Through-Joy organisation, which carry Volkswagen savings stamps. No liability is assumed for payments unless these are made against the immediate exchange of savings stamps in the amount of the payment. At least one stamp in the value of 5 marks is to be posted in the book each week, such stamp to be cancelled by the holder by writing the date of purchase thereon. A special premium must be paid in the case of open cars and cabrio-limousines.

(3) Each car carries limited insurance against collision and public liability for a period of two years from the time the car leaves the plant, the cost of such insurance to be charged against the purchaser.

(4) Payments must be made at the designated places of payment by the purchaser, though collectors may be employed.

(5) When all the spaces in the savings book have been filled with stamps it should be surrendered at the Strength-Through-Joy office in exchange for a new book. After

the start of production an order number will be issued through the nearest District Office. The final savings book is to be turned in at the Strength-Through-Joy office in exchange for a certificate of ownership. Lost or mislaid savings books cannot be replaced.

6) For reasons of technical improvements and the consequent low price of the **Volkswagen**, no interest will be paid on savings.

(7) Until further notice, the **Volkswagen** will be produced in a **deep blue-grey finish**.

(8) Volkswagen contracts are non-cancellable. In exceptional cases the Strength-Through-Joy District office may authorise cancellation. In such cases a fee mounting to 20 per cent of the payments made will be retained.

9) Inquiries should be directed to the nearest office listed in this leaflet.

10) In case of change of residence the savings book must be submitted for correction to the agent who originally issued it. (See **Clause** 1)

11) Tampering with the savings book is punishable by law.

12) Applications for issuance of a savings book may be rejected, without reason.

13) No supplementary agreements to the foregoing rules shall be binding.

14) In case of legal disputes the courts of the city of Berlin shall have the jurisdiction.

-----oOo-----